UNIFORM DRAWING FORMAT MANUAL

NEW CADD AND DRAFTING STANDARDS FOR BUILDING DESIGN AND WORKING DRAWINGS

Fred A. Stitt

McGraw-Hill
New York San Francisco Washington, D.C. Auckland Bogotá
Caracas Lisbon London Madrid Mexico City Milan
Montreal New Delhi San Juan Singapore
Sydney Tokyo Toronto

McGraw-Hill

A Division of The ***McGraw-Hill*** *Companies*

1 2 3 4 5 6 7 8 9 0 DOC/DOC 9 0 9 8 7 6 5 4 3 2 1 0 9

ISBN 0-07-134421-7

The sponsoring editor for this book was Wendy Lochner, and the production supervisior was Pamela A. Pelton.

Printed and bound by R. R. Donnelley and Sons Co.

This book is printed on recycled, acid-free paper containing a minimum of 50% recycled, de-inked fiber.

DEDICATION

Dedicated to the quiet techno heroes of the CSI and the AIA
who strive to bring order and sanity to the process.

ACKNOWLEDGMENTS

Thanks to the crew:
Penny Burbank, Charles Sholten, Chandler Vienneau,
and the ever supportive editorial team at McGraw-Hill,
Wendy Lochner and Robin Gardner.

UNIFORM DRAWING FORMAT HANDBOOK

TABLE OF CONTENTS

SOME HISTORY

From the late 70's through the 90's, your author created a number of systematic working drawing production methods and new document formats suited to computer-aided design and drafting.

The early work waspresented in the books, *Systems Drafting* and *Systems Graphics,* published by McGraw-Hill in the early '80's, and updated and elaborated on in the *Architect's Detail Library and Production Systems for Architects and Designer* published in the early 90's by Van Nostrand Reinhold.

We also created database files that will automate repetitive aspects of all architectural services: design, project management, drawing organization, notation, details, construction administration, and facilities management.

Before long, all the components and stages of architectural service will be one integrated process that will be documented on project management internet sites.

In other words, all A/E documents will be cross-coordinated web sites, with myriad links to a universe of related information, rather than just simple, stand-alone construction documents. They will include a vast array of reference materials (such as codes, reference standards, manufacturers' product literature), and multimedia such as video streams of construction processes.

A UNIFORM DRAWING FORMAT: WHAT IS IT? WHAT GOOD IS IT?

Virtually everyone in the profession is aware of what the Construction Specifications Institute has done in terms of providing a consistent framework and format for writing construction specifications.

Now, when contractors review a set of specifications, they know where to look for any particular set of data -- Sitework is Division 2, Concrete is Division 3, etc. And they can expect each section of the specifications to be written in a certain sequence and include certain data in a format like the following:

The consistent national format helps speed up the bidding and construction processes and provides procedural guidance for specifications writers. It allows all participants in the industry (especially trade associations and product manufacturers), to provide information in a way that fits in content and format with all the other information that design professionals provide in their documents.

It's been a great success, and we've all benefited from it.

Now a grand effort is underway to do the same for working drawings: a set of format and content standards that will bring consistency in working drawings to offices everywhere.

The CSI, with the support and cooperation of the AIA and other private and public agencies, is creating the new system described in the pages that follow. The goal of the new system is to create a basic framework that everyone can use, and still provide flexibility for special cases and office idiosyncrasies.

The basics of the new system are available now, as described in the pages that follow.

THE CSI AND THE UNIFORM DRAWING SYSTEM

The Construction Specifications Institute has a long history of creating and providing documents and education to the profession, to streamline technical communication between architects, consultants, contractors and materials suppliers.

The CSI's best-known achievement is the 16-division Master Format, used by thousands of design offices. The system is the undisputed industry standard for specifications organization and format.

In 1990, to bring the same kind of order to drawings and help coordinate working drawings more closely with specifications, the CSI Technical Committee established the Drawing Subcommittee. Its job was to review drawing/specifications coordination, particularly in the context of computer documentation.

Over one hundred organizations and A/E/C offices were surveyed regarding their systems. Not unexpectedly, there was little in the way of industry-wide standards in drawing formats and organization. Every office and agency had its own standards, which varied widely within offices from project to project, depending on the predilections of individual project managers.

Since 1994, CSI has resolved to create a Uniform Drawing System, to provide consistent national standards for formats, location, sizing, and cross-referencing of drawing components.

Standardization is not a goal for its own sake; it has a purpose. The purpose is to help expedite production, aid in the retrieval and reuse of information, and improve construction communication throughout the industry.

THE ROLE OF THE AIA

The American Institute of Architecture supports these standardization efforts, and the CSI system will include the 1997 version (and subsequent versions) of the "AIA CAD Layering Guidelines." Some systems shown in this book are much simpler than the AIA systems; you may wish to study their total system in depth.

For the latest version of the AIA Layering Guidelines, contact www. AIANET.

Drawing system publications from the CSI to date include:

-- Man*ual of Practice, Construction Documents Fundamentals and Formats Module,* Section FF/010, FF/030, FF/080, and FF/090.

-- *Standard Reference Symbols©, TD-2-6; Project Design Team Coordination and Checklist©, TD-2-7; and Abbreviations©, TD-2-4.*

The CSI can be reached at 1-800-689-2900. Their web site is www. CSINET.org

Your author can be reached at 1-800-634-7779, SFIA@AOL. COM.

THE COMPONENTS OF A UNIFORM SYSTEM

A Uniform Drawing Format provides standards for:

1) Uniform Drawing Set Organization and indexing -- a consistent method of drawing sequencing and numbering.

2) Uniform Drawing Sheet Format -- a modular system of drawing sheet division and subdivision, including a standard format for title blocks.

3) Uniform Detail Format -- a modular system of detail organization that provides the basis for the detail sheet module.

4) Uniform Production Procedures -- a rational sequence of steps and stages for efficient working drawing production.

5) Layering Systems -- separating data on drawing layers or CADD to maximize reuse of repeat data.

6) Filing Systems -- a uniform method of storing and retrieving details and symbols.

7) Uniform Schedule Formats -- the most efficient methods of conveying lists of finishes, doors and windows, hardware, fixtures, and equipment.

8) Uniform Standard Notation and Keynoting Systems.

9) Uniform Production Management Procedures.

UNIFORM DRAWING SET ORGANIZATION

SEQUENCING AND NUMBERING

The CSI has wisely concluded that an alphabetical sheet numbering system is the best standard with which to start.

This system is already familar to most architects and engineers, and it just takes a little tweaking and enrichment to achieves significant advantages over simpler traditional methods.

The content of drawings that includes the major disciplines is the same as aways: Civil Engineering, Landscaping, Structural, Architectural, etc. And each such drawing identifier starts with the first letter of the discipline or consultant: C for Civil, L for Landscaping, and so on.

The basic drawing sheet types don't change either: general information, broadscope plans, narrowscope plans, elevations, cross sections, partial sections, details, and schedules.

Here we need to make some differentiations:

"Construction documents" for a project include contracts, working drawings, clarification drawings, specifications, shop drawings . . . the works. Sometimes they are called "Contract Documents" and sometimes they are issued in different phases or forms, such as "Bidding Documents."

"Working drawings" are specifically the portion of the documents that shows, graphically, how to build the building -- how large things are, what they're made of, and where they are located.

UNIFORM DRAWING SET ORGANIZATION continued

SEQUENCING AND NUMBERING continued

"Sheets" are the individual "pages" of a set of working drawings. A sheet is usually devoted to the data of one discipline, such as architectural, structural, or electrical. It may include one type of information, such as a plan, or it may have a mixture of data: plans, sections, schedules, general notes, etc.

"Drawings" can be any large or small part of a working drawing set or sheet. "Drawing" is the generic term for all graphic information on a sheet. "Drawings" might be or include photos, notes, and graphic information -- hand drawing, reused drawings from a previous project, new computer drawing, photocopies, a construction detail or even a North arrow.

"Layers" are parts of a drawing separated "vertically." Thus you might take a basic floor plan as a "base layer" and combine it with a lighting fixture plan layer to make a complete

"Layers" are parts of a drawing separated "vertically." Thus you might take a basic floor plan as a "base layer" and combine it with a lighting fixture plan layer to make a complete lighting plan. Or the base layer floor plan could be combined with a furniture layer and a fixtures layer to make a fixtures and furnishings plan. The fixtures and furnishings plan might be combined in turn with an electrical power and switching plan.

We all know these things, of course, but we have to make sure we have a common understanding of terminology, as we start to reorganize and standardize the working drawing production process. Some newer, less familiar concepts:

"Models," in computerize, are the components of the building itself. "Sheets" contain the drawing data, such as notes, symbols, title blocks.

"Database" is a systematic organization of data. A database may include project-specific data, such as "models" of the building, and reusable data from a standards database, such as titles, standard notation, details, etc.

"Reports" are an assemly of information from a database. In CADD terms, the "model" is the database for the building, and the "sheets" are database reports of specific aspects of that data.

STEP 1 IN DRAWING SHEET IDENTIFICATION THE FIRST DESIGNATOR

Here is where we "enrich" the traditional alphabetical system. It's a little extra trouble at first, but there are rewards on the other end that make it worth doing.

First, we start with the alphabetical identification of the discipline, in the order they might appear in drawings for a mid-size project.

G General Information for the project and drawing set.

C Civil Engineering.

L Landscaping.

A Architectural.

S Structural Engineering.

M Mechanical -- HVAC.

P Plumbing (or M for Mechanical).

E Electrical Engineering.

T Telecommunications.

Within the site drawings, there might be sheets of Civil, Mechanical, and Electrical work, and this work might well be separated from these disciplines' work within the building itself.

Other drawing sheet types might include:

H Hazardous waste or materials.

F Fire protection.

I Interiors (including fixtures and furnishings).

K Kitchen consultant drawings.

R Reference or Resources.

So how do you identify other specialties, such as Acoustical Drawings? They can't be an A or a C -- they're already taken.

When including additional highly-specialized or hybrid drawings, it might be time to use double designators, such as AC for acoustical.

Notes:

____

____

____

____

____

STEP 2 IN DRAWING SHEET IDENTIFICATION
THE DISCIPLINE DESIGNATOR

On larger or more complex projects, it might be appropriate to use a second alphabetical designator for each drawing type, such as:

ES Electrical - Site.

MP Mechanical - Plumbing.

IF Interior - Furnishings.

The CSI suggests secondary designators, such as the folowing for Telecommunications drawings:

TA Telecommunications - Audio Visual.

TC Telecommunications - Clock and Program.

TI Telecommunications Intercom.

Etc.

Notes:

____

____

____

____

____

____

STEP 3 IN DRAWING SHEET IDENTIFICATION THE DRAWING TYPE DESIGNATOR

After identifying the discipline through a one- or two-letter code, the sheet type comes next.

THE CSI PROPOSAL

The CSI proposes that sheet types can be categorized in one to nine types, so only one number is required, as follows:

0 General Information.
(Mechanical General Information would be M-0, for example.)

1 Plans.

2 Elevations.

3 Sections.

4 Details.

5 Schedules.

6 (open designation)

7 (open designation)

8 Photos or other 3-D representation.

9 (open designation)

This is very close to systems that have been successfully used for years. But it can be improved upon, as you'll see in the next few pages.

WHEN AND WHY HAVE A SHEET TYPE DESIGNATOR?

If you're dealing with 10 or 20 sheets (the majority of working drawing sets), about half the sheets will be architectural, and half will be from consultants.

For a small project, there's not much point in naming architectural drawings anything beyond the list below. And the smallest projects or interior design work may not have separate engineering drawings at all.

Notes:

WORKING DRAWING SHEET IDENTIFICATION continued

SMALL PROJECT SHEET DESIGNATORS

A-G General Information.

C-1 Site plan.
(If there is no civil engineer involved, the site plan is usually called A-1.)

F-1 Foundation plan.
(If there are no structural engineering drawings, this could be A-1 or A-2. Subsequent drawings are numbered A-3, A-4, etc.)

A-1 Floor Plan.

A-2 Exterior Elevations.

A-3 Cross Sections (wall sections may be included).

A-4 Interior Elevations (cabinet work may be included).

A-5 Roof Plan (may be combined with Site Plan).

A-6 Reflected Ceiling Plan
(may be combined with Electrical).

A-7 Door and Window Schedules (may include details).

A-8 Construction Details.

S-1 Foundation and details.

S-2 Framing and details.

M-1 Heating, Ventilation, and Air Conditioning.

M-2 Plumbing.

E-1 Electrical.

DIVISIONS AND SUBDIVISIONS OF ARCHITECTURAL DRAWINGS

If the project is larger and more complex, say multistory with below-grade work, the architectural drawings can be identified as follows:

A.0 General Information, index sheets, symbols, nomenclature, location map, etc.

A.1 Plans.

A.2 Exteror Elevations.

A.3 Cross Sections and Wall Sections.

A.4 Interior Elevations and Finish Schedules.

A.5 Roof Plan.

A.6 Reflected Ceiling Plan.

A.7 Door and Window Schedules.

A.8 Construction Details.

The system in more detail:

A.0

A.0 General Information.

A.1

A-1.0 Lower Level Plan (basement).
(If there are multiple levels below grade, they can be called A-1.0.1, A-1.0.2, etc.)

A-1.1 Ground Floor (first floor) Plan.
(An intermediate level or mezzanine can be designated as A-1.1.2.)

A-1.2 Second Floor Plan.
A-1.3 Third Floor Plan.

A.2

A-2.1 Exterior Elevations.
A-2.2 Exterior Elevations.

A.3

A-3.1 Cross Sections.
A-3.2 Wall Sections.
A-3.3 Wall Construction Details.

A.4

A-4.1 Interior Elevations.
A-4.2 Cabinetwork, Built-in Funishings.
A-4.3 Interior Finish Schedules
(if not with floor plans).

WORKING DRAWING SHEET IDENTIFICATION continued

DIVISIONS AND SUBDIVISIONS OF ARCHITECTURAL DRAWINGS continued

A.5

A-5.1 Roof Plan.

A-5.2 Roof Details.

A.6

A-6.0 Reflected Ceiling Plan -- Lower Level.

A-6.1 Reflected Ceiling Plan -- Ground Floor Level.

A-6.2 Reflected Ceiling Plan -- Second Floor Level.

A.7

A-7.1 Door and Window Schedules (or door only/window only).

A-7.2 Door and Window Frame Schedules.

A-7.3 Door and Window Details (or door only/window only).

A.8

A-8.1 Miscellaneous Construction Details.

A-8.2 Etc.

If using decimal points creates a problem with your CADD or database software, you can use hyphens, like A-1-1 instead of A-1.1. There are other options on the next page.

COORDINATING ARCHITECTURAL SHEETS WITH OTHER DISCIPLINES' DRAWINGS

If the architect's 1.1, 1.2, etc. drawings are floor plans, and the 2.1, 2.2, etc. are elevations, the engineers and other consultants can follow the same pattern.

Thus:

C-0 Civil Engineering General Information.

C-1.1 Civil Engineering Excavation Plan.
C-1.2 Civil Engineering Grading Plan.
C-1.3 Drainage.
C-1.4 Paving.

C-2.1 Profiles, Cut & Fill.
C-2.2 Retaining Walls.

C-3.1 Sections and Details.

Notes:

____

____

____

____

____

WORKING DRAWING SHEET IDENTIFICATION continued

CSI-RECOMMENDED SHEET SEQUENCE DESIGNATOR

There are arguments for naming actual sheet numbers with at least two numbers: 01 instead of 1, 02 instead of 2, etc.

It's suggested that doing so might work better for computer filing and database facilities management.

In that case, here's how the architectural drawing from the previous page could be listed:

A-100 Lower Level Plan (basement).
A-101 Ground Floor Plan.
A-102 Second Floor Plan.
A-103 Third Floor Plan.

A-201 Exterior Elevations.
A-202 Exterior Elevations.

A-301 Cross Sections.
A-302 Wall Sections.
A-303 Wall Construction Details.
Etc.

The zero replaces the . or - between the drawing type and the sheet number. It looks a little better, but it will be confusing at first, and takes some getting used to.

You can eliminate the hyphen if you want to: A101.

WORKING DRAWING SHEET IDENTIFICATION continued

CSI-RECOMMENDED SHEET SEQUENCE DESIGNATOR continued

Or add a space instead of a hyphen, for A 101. The total sheet sequence number takes the following form:

Discipline	Separating Hyphen or space	Drawing Type or Division (plan, elev. etc)	Two-digit Sheet Number
A	-	1	01
A	-	1	02

That's five units in the numbering system. Although there may be over a hundred sheets in a total set of drawings (except for buildings that go over 99 stories), it's unlikely there will be more than 99 sheets for any particular sheet type.

If there are more numbers to add for any "humungous" project, you can create a new category of suffixes and add them to the number.

Notes:

____

____

____

____

WORKING DRAWING SHEET IDENTIFICATION continued

VARIATIONS

The system using hypens and dots as separators:

A-1.1 Lower Level Plan (basement).

A-1.2 Ground FloorPlan.

A-1.3 Second Floor Plan.

A-1.4 Third Floor Plan.

A-2.1 Exterior Elevations.

A-2.2 Exterior Elevations.

The system using double-digit sheet numbers:

A-1.01 Lower Level Plan (basement).

A-1.02 Ground FloorPlan.

A-1.03 Second Floor Plan.

A-1.04 Third Floor Plan.

A-2.01 Exterior Elevations.

A-2.02 Exterior Elevations.

The system without hyphens or dots:

A101 Lower Level Plan (basement).

A102 Ground FloorPlan.

A103 Second Floor Plan.

A104 Third Floor Plan.

A201 Exterior Elevations.

A202 Exterior Elevations.

The variations shown above are all readable, so long as they're clearly explained to users. Your choice is a matter of personal judgment and possible computer limitations.

FILES -- ELECTRONIC AND OTHERWISE

According to the CSI system, the main file types are "project" and "library."

"Library" files are reusable information and graphics like standard construction details or standard notation. Symbol libraries, in particular, are almost universally used as part of CADD production.

"Project" files, not surprisingly, are project specific.

"LIBRARY" FILES

Here are the primary **library** file categories:

Symbols (such as north arrows, drawing titles, door and window symbols, and plumbing, electrical, and HVAC symbols).

Standard Construction Details.

Schedule Formats (and sometimes content).

Standard notation and keynotes.

Production Management such as:

-- Sheet Files -- formats and title blocks.

-- Standard drawing indexes.

-- General information sheet data.

-- Formats for working drawing mockup sets.

-- Layer management codes.

Symbols can be catalogued by division of work: General Information (where symbols and materials indications are usually explained), Sitework, Architectural, Electrical, Plumbing, etc.

Standard Construction Details are best filed according to CSI coordination number. Thus sitework detail files will start with the division number 2 (or 02), paving is 025, and concrete paving is 28, for a final number of 02528.

While details require a clearly differentiated file numbering system, they're easiest to find and sort through an easily understood locator system, such as alphabetical. Thus all stair detail files -- whether wood, concrete, or metal -- could be located under S for "Stairs." Windows would be under W, Doors under D, etc. These file copies should be on paper, in 3-ring binders, to make them as accessible and easy to use as possible.

Using the simplest and most direct locator system helps assure that such files will be used. If the final file number is also the locator number, it' will slow people down, and they'll avoid the system.

Standard Schedule formats are readily located by type and discipline, such as Finish, Door, Window, Hardware, Cabinet, etc. Specific file numbers would be CSI number correlated, as with details.

Standard Notation is most readily clustered and organized by drawing types. Thus, you might have notes for sitework that includes CSI divisions sitework (02), concrete (03), masonry (04), plumbing (015), and electrical (16). Rather than finding sitework notes one by one through each separate division file of notation, a more convenient file would include clusters of notes from all appropriate divisions for each sheet type.

"PROJECT" LIBRARIES

The primary **project** file categories are:

Construction Details -- original, edited standard details, and unedited standard details. These are file-numbered according to their drawing, and, possibly, their sheet coordinate location numbers on the drawing.

Each such drawing should also include its **Library File Number** for cross reference and easy back referencing. See examples in the chapter on Construction Detail formats.

Project-specific **General Information**. Combine with standard General Information and locate by sheet coordinate numbers.

Project-specific **Notation**. This will probably just be combined with the standard notes and keynotes, with no special designator.

Project-specific title block and other **Production Management** information. Some of this may be linked with data base files and project management files.

Database and embedded or linked information. A wall symbol may reference a specific type of wall construction that is shown in detail. The detail includes notation that refers to specifications. The specification data may link with manufacturer or trade association information, building code requirements, and construction cost reference information. The complete links may become standards as they're created for new projects, and then be assigned to Library files.

With greater use of the internet and linked construction file information, database information will be much more dynamic.

It may be standard, but its links to other, related data, will be dynamic and no longer limited to a sheet of paper on a table in the construction shack.

Shop drawings. Every office has its way of processing and filing shop drawings; since they are a combination of electronic and paper, graphics and text, there's little consistency in the actual form of the information. We'll leave it to the CSI to standardize the identification and filing of these documents.

Notes:

UNIFORM DRAWING SHEET FORMAT

REASONS AND ADVANTAGES

The rationale for a national standard for drawing numbering and sheet organzation is the same as the idea behind the CSI specifications Masterformat: Consistency in format helps everyone know where to put data and where to find it.

In terms of working drawings, such consistency makes it easier to create, organize, cost, bid, and build.

It will be easier for manufacturers to provide detail drawings A/E's can directly reuse in their documents. Details can be as consistent in size and appearance as other design professionals' drawings.

Consistent file numbering will make it easier to find resuable data from previous projects.

As the internet becomes architects' and engineers' daily working tool, it will make it far easier to exchange data electronically.

As A/E's use more independent contractors to assemble projects, or as offices team up at far distances, it will be easier to plan, coordinate, and assemble their work according to a rational, preset pattern.

It's all a natural "next step" for the CSI, and it has the full support of the AIA. As a corollary, government agencies at all levels, including local building departments, will inevitably get on board and try to turn what is a fundamentally good idea into a complex set of regulations. (We'll see if we can persuade them not to do that.)

A MODULAR SYSTEM FOR SUBDIVIDING THE DRAWING SHEET

There are several reasons: Why create a consistent grid or module as a part of the design of every working drawing sheet?

1) It allows for orderly positioning of construction details and other elements, so that everything lines up neatly.

2) It provides a "locator" system for finding and communicating information (like the alphanumeric coordinates on a road map).

3) It requires a consistent system of organizing the data in individual construction detail drawings, which enhances clarity and expedites the creation and use of standard detail libraries.

4) It will provide numbered Drawing Blocks that can be readily identified in your CADD system.

Many years ago, your author worked out the construction detail window that is the basis for the working drawing sheet module. It's the optimal size for all construction details when they are drawn at their most appropriate scales and sizes.

Besides being the best size for construction details, the module also had to be a reasonable subdivision for the varied sheet sizes used by the profession. Ultimately, we have to tweak a bit, to allow for varied sheet margins and title blocks. But when fudging, just allowing a little extra space for the detail modules is not a problem;

More on the inner workings of detail drawings, The next chapter but for now, here is the recommended module -- the first step in putting together a national Standard Detail Library.

UNIFORM DRAWING SHEET FORMAT continued

MODULAR SYSTEM FOR DRAWING SHEET DIVISION continued

This sample is for a 24" x 36" drawing sheet size, the size used for about 90% of all working drawings.

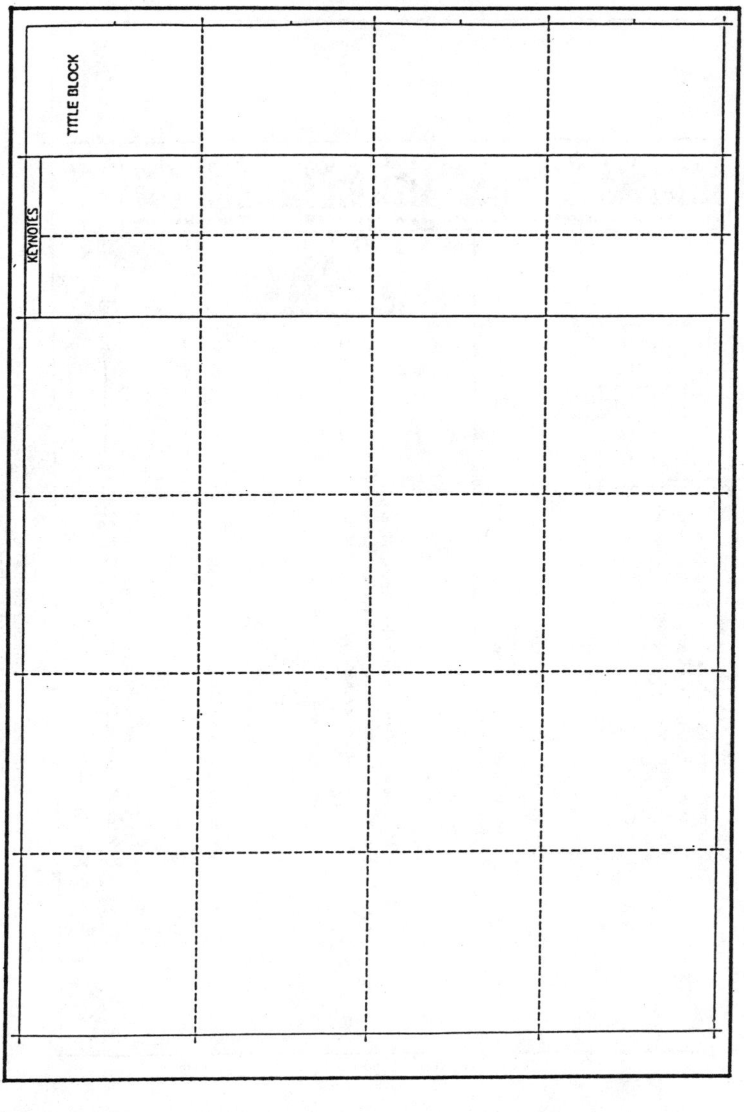

TITLE BLOCK

KEYNOTES

UNIFORM DRAWING SHEET FORMAT continued

MODULAR SYSTEM FOR DRAWING SHEET DIVISION continued

The 24" x 36" drawing sheet showing keynotes and other data in relationship to the uniform modular system..

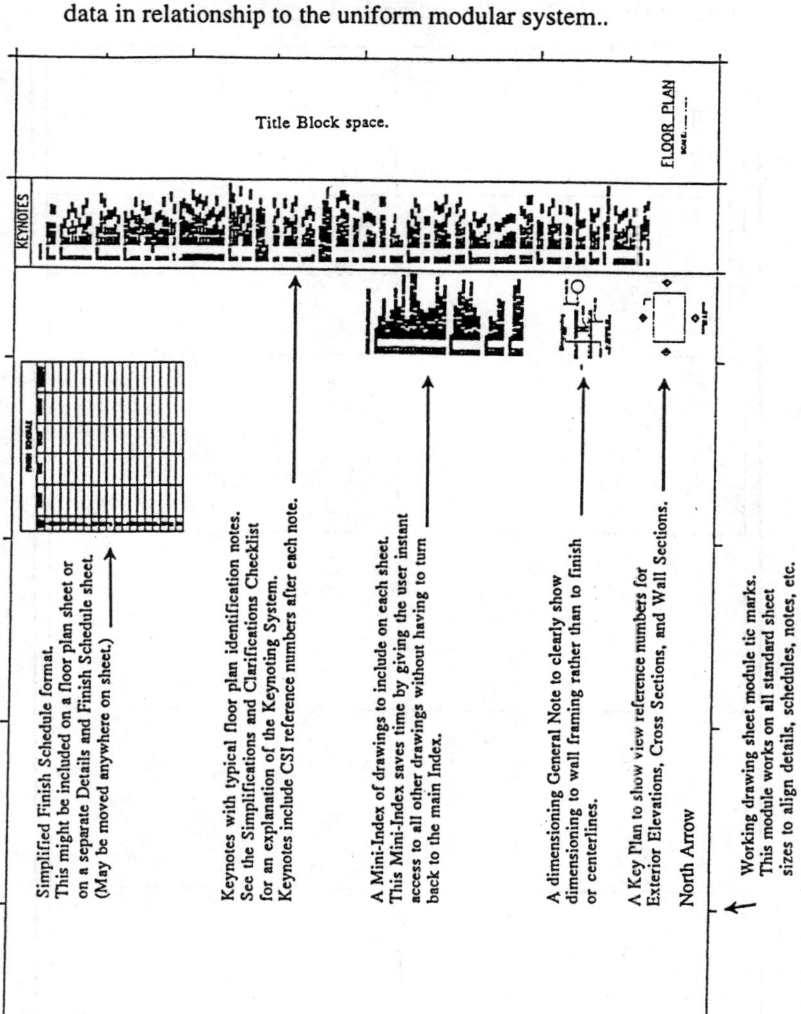

UNIFORM DRAWING SHEET FORMAT continued

MODULAR SYSTEM FOR DRAWING SHEET DIVISION continued

The uniform drawing module system with a 30" x 42" drawing sheet.

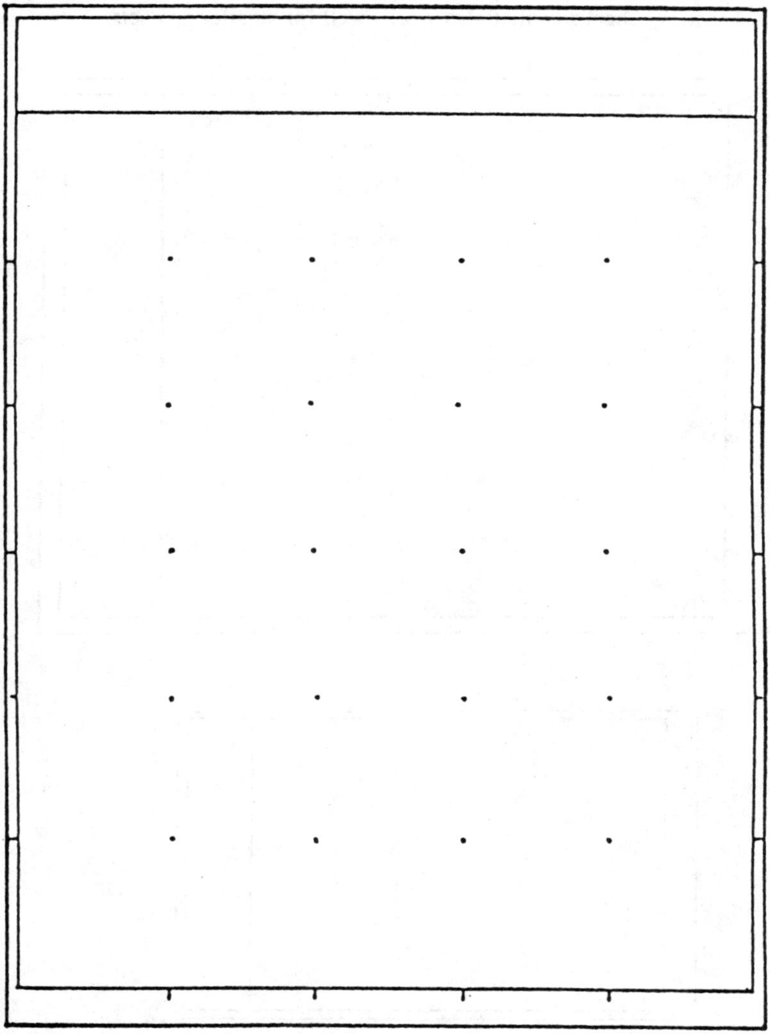

UNIFORM DRAWING SHEET FORMAT continued

MODULAR SYSTEM FOR DRAWING SHEET DIVISION continued

The uniform drawing module system with an 18" x 24" drawing sheet. This sheet size is used for smaller work and remodeling jobs. At the bottom of the page is a sample of the 11" x 17" sheet size that also works well for smaller projects.

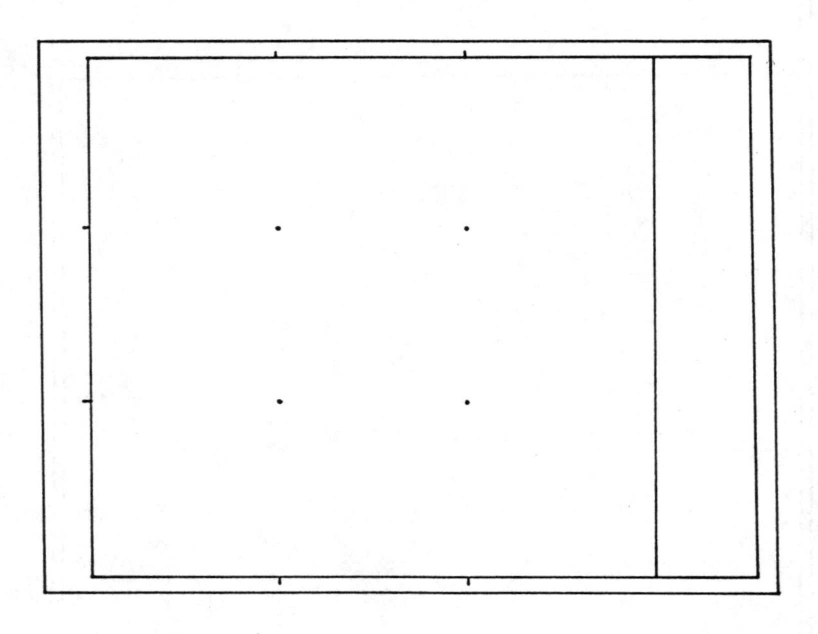

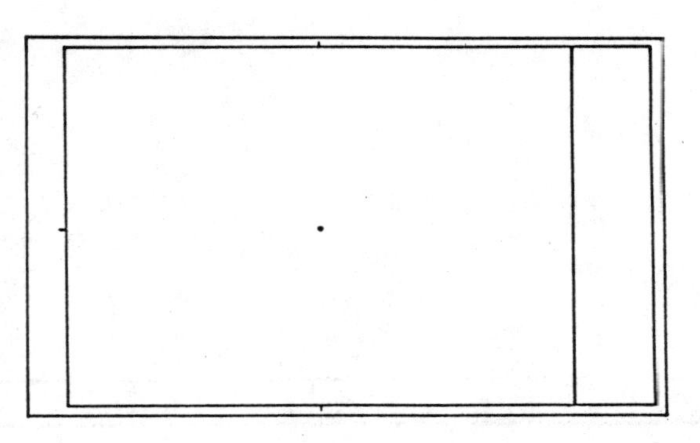

UNIFORM DRAWING SHEET FORMAT continued

MODULAR SYSTEM FOR DRAWING SHEET DIVISION continued

The recommended format for 8-1/2" x 11" supplemental sheets.

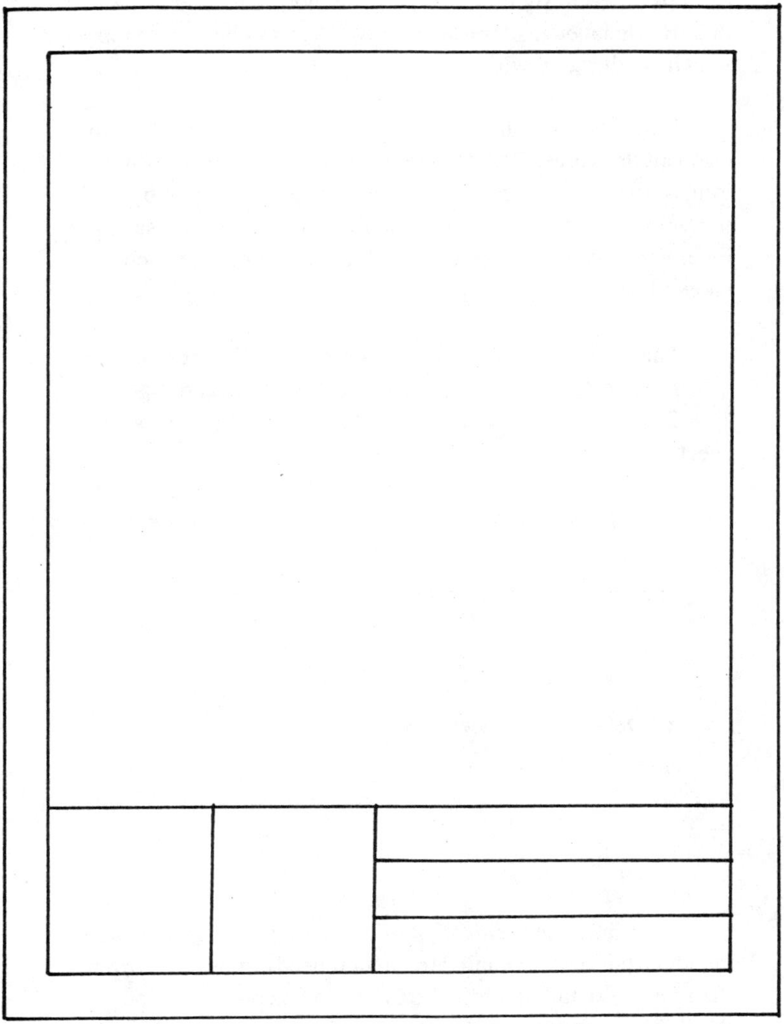

THE DRAWING SUBDIVISION COORDINATES

The CSI recommends that the numerical coordinates proceed from left to right, with which we thoroughly agree. And that the alphabetical coordinates proceed from bottom to top, which we disagree with.

If the CSI prevails, thereby mandating a standard for government drawings, CADD systems, etc., then so be it. But as long as there's a choice, and it doesn't cause confusion by dealing with other documents using a different system, we strongly recommend using a "top-down," alphabetical system such as shown below.

Then any time you need to refer to a drawing component, schedule, or detail location, you can call it out accordingly as 1-1, 4-C, etc. Keynotes would always start at 6-A on this size sheet.

Should you use this coordinate system for identifying details?

The traditional method lists a detail as something like:

1

A-3

for Detail #1 on Sheet A-3.

Now it might be:

1-A

A-3

Is this an improvement? Offhand, it just seems to make the identifier more complicated than it needs to be. Other ramifications of detail identification need to be considered too. We'll look at the pros and cons in the next chapter.

SUPPLEMENTAL DRAWINGS

Supplemental information may be provided on 8-1/2" x 11" sheets.

Margins are 1/2 inch top, right hand side, and bottom; 3/4" at left or binding margin.

These sheets may be issued as horizontal or vertical and include a mini-title block as the bottom strip or right hand strip. There are not particular recommendations for information or format except that the following information is required:

__ Design logal, design firm name.

__ Stamp space may be reuqired.

__ Project identification.

__ Supplemental sheet name.

__ Supplemental sheet number.

__ Supplemental sheet management:

Date, reference, drawn by, checked by, etc.

UNIFORM DRAWING SHEET FORMAT continued

TYPICAL REFERENCE INFORMATION ON EACH SHEET

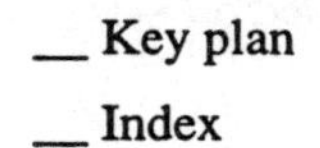

__ Key plan

__ Index

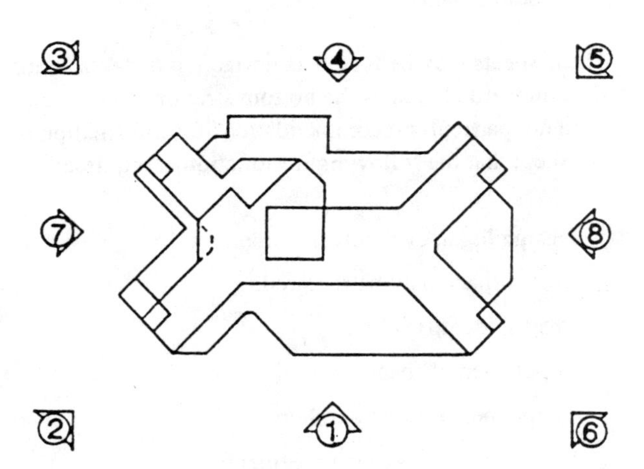

DRAWING CROSS-REFERENCE INDEX

ARCHITECTURAL

A0.1	PERSPECTIVE
A0.2	GENERAL INFORMATION
A1.1	SITE PLAN
A1.2	IRRIGATION PLAN
A1.3	LANDSCAPING/PLANTING PLAN
A1.4	SITEWORK DETAILS
A2.1	FLOOR PLANS & FINISH SCHEDULE
A2.2	ROOF PLAN
A2.3	FLOOR PLANS, DOOR/WINDOWSCHEDULES
A2.4	DETAIL PLANS & INTERIOR ELEVATIONS
A3.1	EXTERIOR ELEVATIONS
A3.2	CROSS SECTIONS
A3.3	WALL SECTIONS
A4.1	EXTERIOR DETAILS
A4.2	EXTERIOR DETAILS
A5.1	REFLECTED CEILING PLANS & CLG. DETAILS
A5.2	STAIRS & MISC. DETAILS
A5.3	CABINETS & INTERIOR DETAILS

STRUCTURAL

S.1	FOUNDATION PLANS
S.2	FOOTING & SLAB DETAILS
S.3	FLOOR FRAMING PLAN & DETAILS
S.4	WALL FRAMING DETAILS
S.5	ROOF FRAMING PLAN & DETAILS
S.6	TRUSS ELEVATIONS
S.7	ROOF DETAILS

MECHANICAL

M1.1	SITE UTILITIES
M2.1	HVAC PLAN
M3.1	PLUMBING FLOOR PLAN
M3.2	PLUMBING ISOMETRICS

ELECTRICAL

E1.1	SITE & LANDSCAPING LIGHTING
E2.1	CEILING FIXTURE PLAN & DETAILS
E2.2	ELECTRICAL POWER PLAN

THE STANDARD DETAIL DRAWING FORMAT

DRAWING SHEET MODULES

COORDINATION WITH STANDARD DETAIL FORMAT

Construction detail drawings and formats must work graphically in three ways:

1) The details must fit within a standardized "window," so they will relate well to each other on larger drawing sheets.

2) The detail window must accommodate details of all types at their most readable sizes and scales.

3) The detail window must work well as a subdivision of the most commonly used drawing sheet sizes.

WHY USE THE 6" X 5 3/4" SHEET MODULE?

It's not an arbitrary size.

It's the best possible size for virtually all construction detail drawings. Here's how your author settled on this module.

1) First, we established the clearest scales for construction details: mainly 1-1/2" = 1'-0" for the simplest details, such as curbs and other minor components in sitework; and 3" = 1'-0" for more complex details that show some combinations of materials windows and walls, skylights and roofs, etc.

2) Then we established the most readable and consistent positioning of detail drawing, notation, and dimensions:

THE STANDARD DETAIL SHEET FORMAT continued

WHY USE THE 6" X 5 3/4" SHEET MODULE? continued

mainly notation on the right, dimensions on the left. This pattern within the detail drawing module brings uniformity and exceptions in positioning are encouraged wherever it helps clarity.

3) Consistency in scales, drawing sizes, positioning, was designed to facilitate merging of standard components in new detail drawings. Standard wall construction details would be at a size that could readily be combined with standard window types; standard roofing sections could be merged with skylights; parapet drawings would with with standard flashing drawings; etc.

4) Finally we set a reasonable (not too large, not too small) standard for the size and position of detail titles, scales, and standard detail file numbers.

Ultimately we achieved a consistant, universal standard: Detail drawings at the most appropriate scales and sizes worked best within a space 6" wide by 5-3/4" high.

This project required weeks of measurement, computation, and comparison with construction details we had compiled from offices across the country. It had to be right, because we were introducing the nation's first national standard construction detail system in book and software form, the Guidelines Standard Detail Library and in book form, *The Architect's Detail Library* published by Van Nostarnd-Reinhold. Fortunately it worked and is now a de facto national standard.

Standard detail window cut lines define a drawing area size of 6" wide by 5-3/4" high. (There is flexibility in this module, so minimum detail window area can be 5-1/2" wide by 5-1/4" high.) The detail window and its features are illustrated on the pages that follow.

THE STANDARD DETAIL SHEET FORMAT continued

THE STANDARD DETAIL FILE SHEET

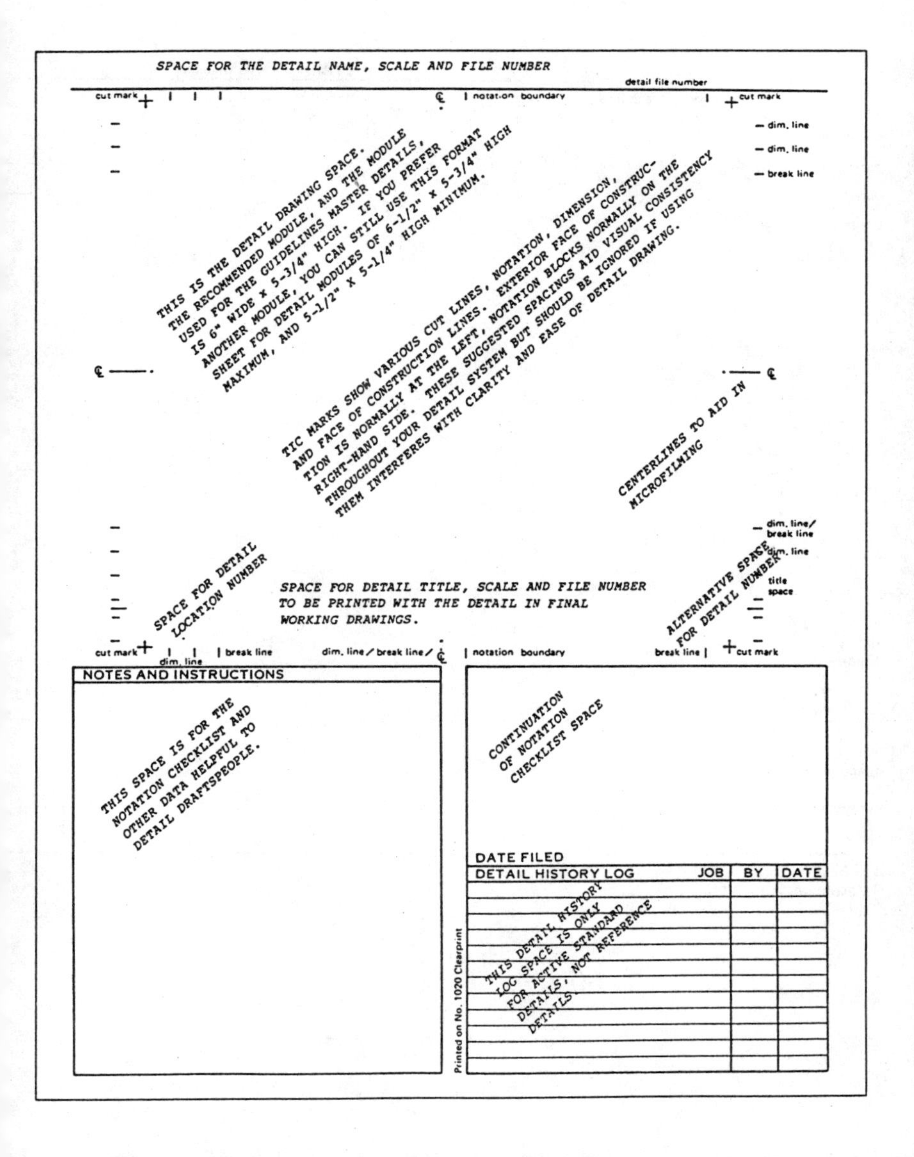

THE STANDARD DETAIL SHEET FORMAT continued

THE DETAIL WINDOW continued

THE DETAIL "WINDOW"

DETAIL FILE NUMBER:

dim. lines
break line
notation boundary
break line
cut mark
dim. line
dim. line
break line
dim. line/
break line
dim. line
title
space
cut mark

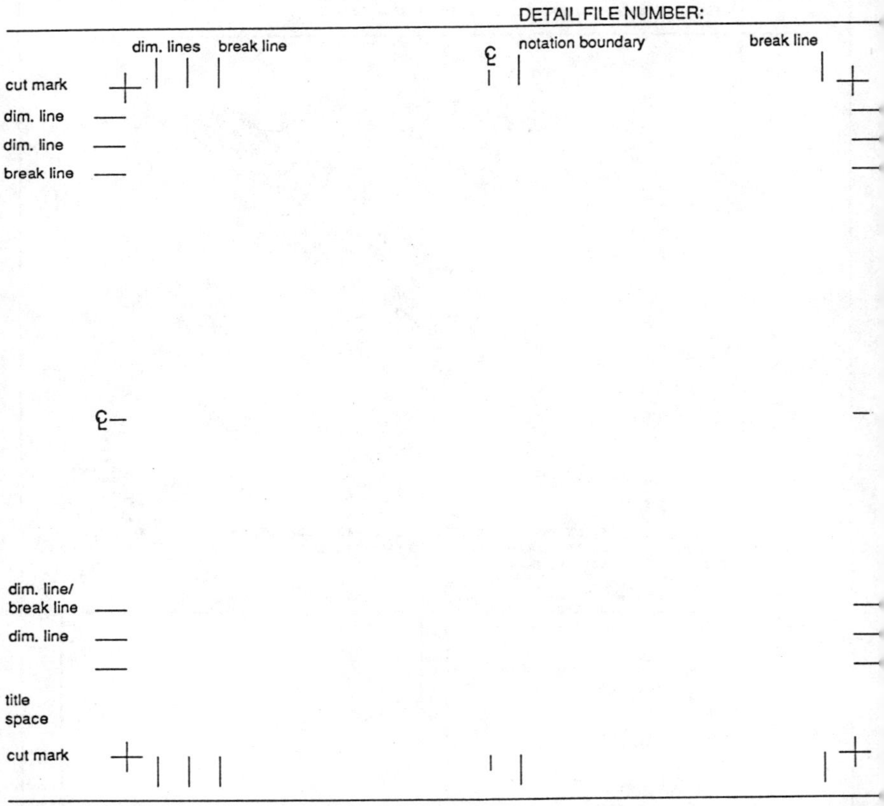

DETAIL INFORMATION
References, jobsite feedback, job history

THE STANDARD DETAIL SHEET FORMAT continued

THE DETAIL WINDOW continued

A DETAIL WITHIN THE "WINDOW"

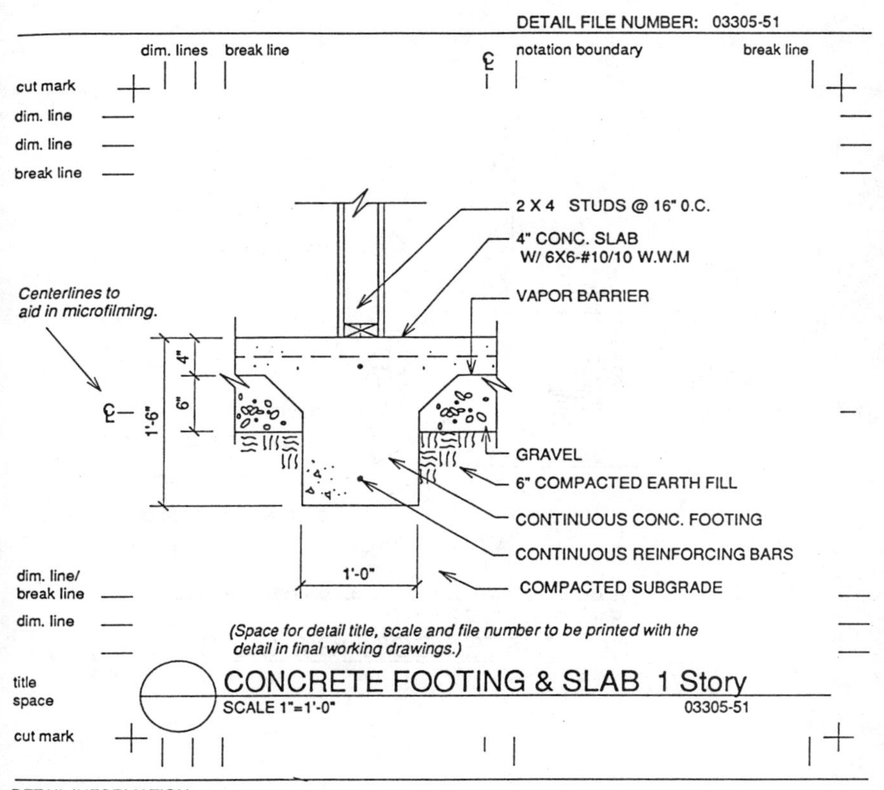

DETAIL INFORMATION
References, jobsite feedback, job history

THE STANDARD DETAIL SHEET FORMAT continued

THE DETAIL WINDOW continued

THE DETAIL AS IT APPEARS WITHOUT THE "WINDOW"

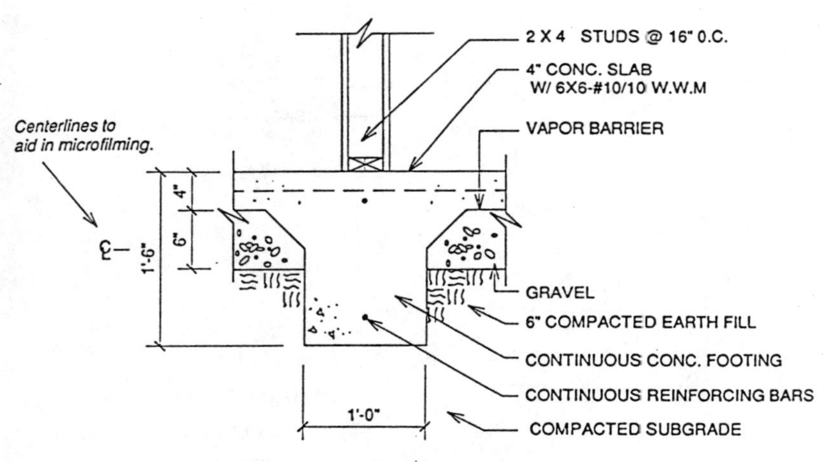

(Space for detail title, scale and file number to be printed with the detail in final working drawings.)

CONCRETE FOOTING & SLAB 1 Story

SCALE 1"=1'-0" 03305-51

THE STANDARD DETAIL SHEET FORMAT continued

THE DETAIL WINDOW continued

ASSEMBLED DETAILS

An illustration of an assembled detail sheet using consistently formatted graphics and notation.

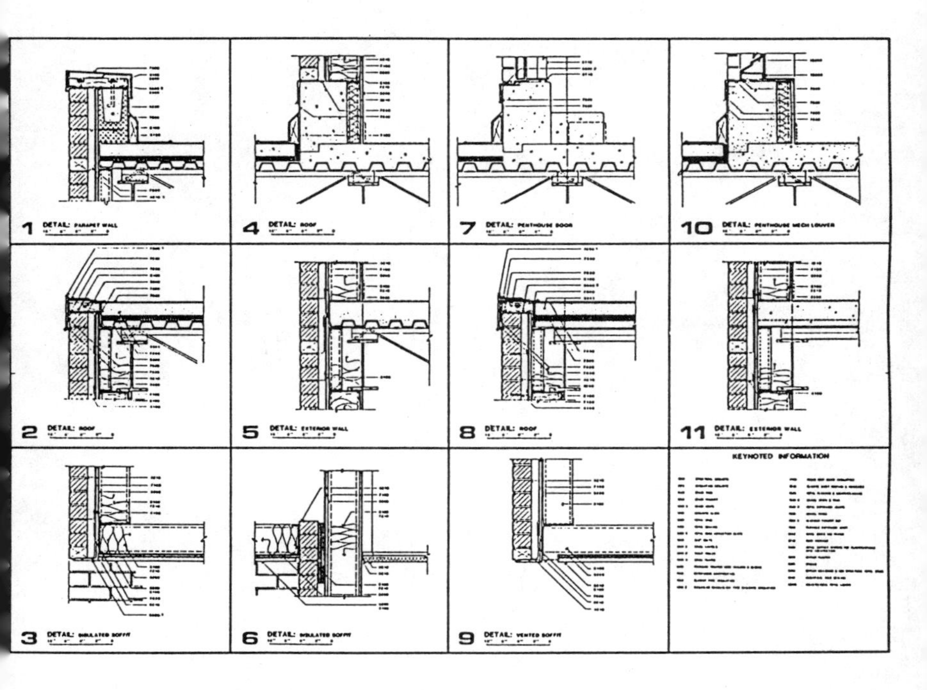

UNIFORM LAYERING SYSTEMS

LAYERS -- WHAT GOOD ARE THEY?

All CADD systems provide for layered information in drawings. The layers were originally a means to control pen plotters -- one layer for one thickness of ink line, another layer for another.

In architectural practice, the value of layering in drafting has been long appreciated as a means of separating different aspects of work and different disciplines and trades.

Thus mechanical and electrical engineering drawings could be created "on top" of a base layer created by the architect -- a significant improvement over the archaic practice of having engineering drafters redraw architectural plans from scratch. Then they could do their work atop an architectural reference sheet while architectural drafters continued with their data atop the same base layers.

Layering facilitates the storage and retrieval of reusable information. A single layer showing sheet border and title block, for example, can be combined over and over with other drawings in a set.

In addition, layering allows output showing one discipline's work as subdued, shadow-print background information in relationship to another's work -- a significant clarifier.

Common ways to identify layers include abbreviated drawing names, letter codes, or numbers. A unified system will help avoid confusion, excessive laying and loss of data on misplaced or misnamed layers.

THE DIFFERENCES BETWEEN FILE NAMES AND LAYER NAMES

The two types of CADD files are sheet files and model files. The differentiation isn't relevant to most smaller projects, but becomes important for managing big jobs.

Model files are the building components, drawn at full scale. They'll be referenced and recombined in other files to make sheet files.

Sheet files are images of building components drawn to scale and combined with drawing reference information such as titles.

DON'T INCLUDE JOB ID NAMES OR NUMBERS

A job number that becomes a permanent part of model or sheet files adds unnecessarily to the length of the file name and can become a headache letter when reusing documents in other circumstances, such as facilities management.

Instead, use folders (Macintosh) or directories (PC's), to link model files and sheet files to their originating projects.

THE AIA GUIDELINES

The AIA's CADD *Layer Guidelines* originally offered two formats for layer names, one using 6 to 16 characters, and a short form using 3 to 8 characters. Since then, the shorter format has fallen into disfavor, and the longer format is considered easier to understand and use.

The AIA system subdivides its layer codes into three parts: "major" group, "minor" group, and "modifier." Each group is separated by a hyphen.

The "major" group designator is the design discipline: architectural, engineering, or consultant. The discipline groups defined in the AIA system are the same as the CSI Uniform Drawing System:

G General Information for the project and drawing set.
C Civil Engineering.
L Landscaping.
A Architectural.
S Structural Engineering.
M Mechanical -- HVAC.
P Plumbing (or M for Mechanical).
E Electrical Engineering.
T Telecommunications.

Site drawings might include sheets of Civil, Mechanical, and Electrical work. This work might well be separated from these disciplines' work within the building itself.

Other drawing sheet types include:

Q Equipment.
H Hazardous waste or materials.
F Fire protection.
I Interiors (including fixtures and furnishings).
K Kitchen consultant drawings.
R Reference or Resources.
X Other disciplines.
Z Contractor or shop drawings.

NUMBERED LAYERS

The layers that follow can be few or many. Layer subdivisions for floor plans might be:

Layer 01: Line drawings

- 01.1 Structural module (S drawing).
- 01.2 Wall/partition module.
- 01.3 Exterior walls.
- 01.4 Interior structural walls and columns.
- 01.5 Interior partitions.

Layer 02: Furnishings, Fixtures and Equipment

- 02.1
- 02.2
- 02.3

Layer 03: Dimensions.

- 03.1 Exterior dimensions.
- 03.2 Interior dimensions.

Layer 04: Notation and note leader lines, or keynotes.

Layer 05: Symbols -- doors.

Layer 06: Symbols -- windows.

Layer 07: Detail and Section keys.

Layer 08: Interior elevation keys.

Layer 09: Drawing title.

Layer 10: Sheet border, title block.

Layer 11: Engineering discipline line drawing.

Layer 12: Engineering discipline dimensions.

Layer 13: Engineering discipline Notation, note leader lines, keynotes.

Layers 14 and up: Additional drawings, schedules, text.

Each layer identifier will start with the discipline code letter -- just like sheet identifiers such as

A Architectural.
S Structural Engineering.
M Mechanical -- HVAC.
P Plumbing (or M for Mechanical).
E Electrical Engineering.

Layer A-01.3 then is Architectural exterior walls.

The problem is that an alphanumeric system like this is hard to remember. A uniform standard for general use should have an alphabetical key for discipline and some logical subsequent identification that people can immediately relate to the information conveyed.

A naming system such as that described on the next few pages provides word codes that simplify information identification.

NAMING LAYERS VERSUS NUMBERING

In contrast to the CSI specifications MasterFormat, which is totally numeric, it makes sense to name (rather than number) drawing layers. Names or abbreviations are easier to remember, and names provide greater numbers of possible options and combinations than a preset numbering system.

Thus an architectural plan starts with the major group designator "**A** ," followed by a hyphen separator. Then its subject, the "minor" group (construction assembly or drawing component), such as walls, doors dimensions, etc., follows with abbreviations:

A-WA

A-WI

A-DO (or DR)

A-DI

The major group is "**A**" for architecture.

SHORT FORM AND LONG FORM

The secondary or minor group uses two letters. Two are required, because there are many subgroup items that start with the same letter, such as "D," which could stand for doors or dimensions. "W" could be walls or windows. So the second letter readily identifies the particular component.

If you use a longer form, then A-WA would be A-WALL, A-WI would be A-WIND, A-DR would be A-DOOR. All of these are easier to identify. The short form was suggested in the first edition of the AIA *Cad Layer Guidelines*, but they've had second thoughts.

UNIFORM LAYERING SYSTEMS continued

SHORT FORM AND LONG FORM continued

The Second Edition of the AIA *Cad Layer Guidelines* recommends dispensing with the short form, the long form being easiest to understand and remember.

A-WALL
A-WIND
A-DOOR
A-DIMN

After the major and minor groups, a **modifier** -- an optional field for differentiating among "minor" groups -- can follow:

A-WALL-BEAR	for bearing or structural wall.
A-WALL-PART	for non-bearing partition.
A-WALL-MATS	wall materials indication

The modifier is optional, a possible clarifier in some instances, but not necessary in all cases.

How many modifiers and layers are too many?

Layer enthusiasts at one time were subdividing drawings down to the door jambs. Others were insisting you only needed three or four basic layers at most.

It all depends on the size and complexity of the project. Basically the more people involved, and the more uses you have for different drawings, the more layers you need.

Until we get a large amount of accumulated experience on a variety of building types, the decisions are judgement calls.

You might need doors and door swings on a floor plan and the door symbols on a separate layer. Or doors might be on a layer separate from walls and partitions and symbols on a third. There are circumstances where one or the other would be most appropriate. But this points up the importance of planning out the documents and clarifying their uses in advance of design development drawings and production.

The AIA system then allows for an additional field defined by the user. This might differentiate one part of a contract from another, or the floor or level of a building such as

A-WALL-PART-01 non-bearing partition -- 1st floor.

A-WALL-PART-01 non-bearing partition -- 2nd floor.

In short form the first one above would be AWL-01.

Besides construction elements, the types of the drawings also have to be identified, such as: plans, elevations, sections, details, schedules AND the type of drawing information such as notes, dimensions, detail symbols . . .

The AIA system designates drawing types as:

PLAN or PL

ELEV or EL

SECT or SE

DETL or DE

For a notation layer, you'd have:
A-WALL-PART-PLAN- NOTES

A BASIC LAYER IDENTIFICATION SYSTEM

See the Second Edition of the AIA CAD Layer Guidelines for their complete list of recommended layer names. (Order through the bookstore at AIAonline.org.) The AIA list is quite extensive and most users won't need as exhaustive a number of layers as their list. But they're there and prenamed for you if you need them.

Here is an adapted system of layers suited to mid-size multi-family housing or two-story commercial building.

DWG #	SHEET TITLE	LAYERS	LAYER CONTENTS
A-00	**General Information**	A-GNOT	Standard General Notes.
		A-SYMB	Standard Symbols Legend.
		A-INDX	Dwg. Index, vicinity map.
		A-NOTE-00	Notes, drawing title & #.
		A-CODE	Code compliance notes.
		A-BRDR-TITL	Sheet border/title block.
C-1.1	**Sitework**	C-PROP	Existing property lines, contours, & grades.
		C-TOPO	New contours, grading.
		A-PLAN	Building plan on site
		A-PAVE	Paving, parking, walks.
		A-APPR	Appurtenances, fixtures.
		C-GNOT	Sitework general notes.
		C-SITE-DIMS	Dimensions.
		C-NOTE	Notes, det. keys, dwg title & #.
		A-BRDR-TITL	Sheet border/title block.
L-1.1	**Landscaping**	L-PLNT	Landscaping plan & notes.
		P-IRRG	Landscape sprinkler.
		C-PROP	Existing property lines, contours, and grades.
		C-TOPO	New contours, grading.
		A-PAVE	Paving, parking, walks.
		A-APPR	Appurtenances, fixtures,
		L-NOTE	Notes, det. keys, dwg title & #.
		A-BRDR-TITL	Sheet border/title block.

UNIFORM LAYERING SYSTEMS continued

A BASIC LAYER IDENTIFICATION SYSTEM continued

DWG #	SHEET TITLE	LAYERS	LAYER CONTENTS
A-1.1	**1st Floor Plan**	A-WALL-PLAN-01	Walls & columns.
		A-WALL-DIMS-01	Dimensions.
		A-WALL-MATS-01	Materials, hatching.
		A-DOOR-01	Doors.
		A-DOOR-SYMB-01	Door symbols.
		A-WIND-01	Windows.
		A-WIND-SYMB-01	Window symbols.
		A-EQUIP-01	Equipment & fixtures.
		A-NOTE-01	Notes, det. keys, dwg title & #.
		A-ROOM-01	Room names, finishes key.
		A-FNSH-01	Room finish schedule.
		A-BRDR-TITL	Sheet border/title block.
A-1.2	**2nd Floor Plan**	A-WALL-PLAN-02	Walls & columns.
		A-WALL-DIMS-02	Dimensions.
		Etc.	
A-2.1	**Exterior Elevations**	A-ELEV-OTLN	Outline of building.
		A-ELEV-MATS	Materials, hatching.
		A-ELEV-DIMS	Dimensioning.
		A-ELEV-NOTE	Notes, det. keys, dwg title & #.
		A-BRDR-TITL	Sheet border/title block.
A-3.1	**Cross Sections**	A-SECT-OTLN	Outline of building.
		A-SECT-MATS	Materials, hatching.
		A-SECT-DIMS	Dimensioning.
		A-SECT-NOTE	Notation & drawing title.
		A-BRDR-TITL	Sheet border/title block.
A-3.2	**Wall Sections**	A-WSEC-OTLN	Outline of wall sections
		A-WSEC-MATS	Materials, hatching.
		A-WSEC-DIMS	Dimensioning.
		A-WSEC-NOTE	Notes, drawing title & #.
		A-BRDR-TITL	Sheet border/title block.

UNIFORM LAYERING SYSTEMS continued

A BASIC LAYER IDENTIFICATION SYSTEM continued

DWG #	SHEET TITLE	LAYERS	LAYER CONTENTS
A-3.3	**Wall Details**	A-WDET-DETS	Wall detail outlines.
		A-WDET-MATS	Materials, hatching.
		A-WDET-DIMS	Detail dimensions.
		A-WDET-NOTE	Notes, drawing title & #.
		A-BRDR-TITL	Sheet border/title block.
A-4	**Interior Elevations**	I-ELEV-OTLN	Outline of walls.
		I-ELEV-FNSH	Materials, hatching.
		I-ELEV-DIMS	Dimensioning.
		I-ELEV-NOTE	Notes, det. keys, dwg title & #.
		A-BRDR-TITL	Sheet border/title block.
A-5	**Roof Plan**	A-ROOF-OTLN	Roof outline.
		A-ROOF-MATS	Materials, hatching.
		A-ROOF-DIMS	Dimensioning, slopes.
		A-ROOF-NOTE	Notes, det. keys, dwg title & #.
		A-BRDR-TITL	Sheet border/title block.
A-5.1	**Roof Details**	A-ROOF-DETS	Roof construction details.
		A-RDET-NOTE	Notes, drawing title & #.
		A-RDET-DIMS	Detail dimensions.
		A-BRDR-TITL	Sheet border/title block.
A-6.1	**Reflected Ceiling Plan -- 1st Floor**	A-WALL-PLAN-01	Walls & columns.
		A-CLNG-01	Ceiling components, soffits, beams, slopes, etc.
		A-LITE-01	Light fixtures.
		A-CLNG-DIMS	Dimensioning, slopes.
		A-CLNG-NOTE	Notes, det. keys, dwg title & #.
		A-BRDR-TITL	Sheet border/title block.
A-6.2	**Reflected Ceiling Plan -- 2nd Floor**	A-WALL-PLAN-02	Walls & columns.
		A-CLNG-02	Ceiling information.
		A-LITE-02	Light fixtures.
		Etc.	

UNIFORM LAYERING SYSTEMS continued

A BASIC LAYER IDENTIFICATION SYSTEM continued

DWG #	SHEET TITLE	LAYERS	LAYER CONTENTS
A-7.1	**Door Schedule & Details**	A-DOOR-FRME	Door frame schedule.
		A-DOOR-SCHD	Door types.
		A-DOOR-HDWR	Door hardware schedule.
		A-DOOR-DETS	Door frame detail dwgs.
		A-DOOR-NOTE	Notes, det. keys, dwg title & #.
		A-BRDR-TITL	Sheet border/title block.
A-7.2	**Window Schedule & Details**	A-WIND-FRME	Window frame schedule.
		A-WIND-SCHD	Window types.
		A-WIND-HDWR	Window hardware sched.
		A-WIND-DETS	Window detail drawings.
		A-WIND-NOTE	Notes, det. keys, dwg title & #.
		A-BRDR-TITL	Sheet border/title block.
A-8.1	**Misc. Construction Details**	A-ACCS-DETS	Accessibility details.
		A-FIXT.DETS	Wall-hung fixture details.
		A-STAR-DETS	Stair details.
		A-CABT-DETS	Cabinet details.
		A-DET-NOTES	Notes, drawing title & #.
		A-BRDR-TITL	Sheet border/title block.
A-8.2	**Misc. Construction Details**	**Etc.**	
M-1.1	**Mechanical -- 1st Floor HVAC**	A-WALL-PLAN-01	Walls & columns.
		A-EQUIP-01	Equipment & fixtures.
		M-HVAC-DUCT-01	Mech. room & air ducts.
		M-HVAC-DIMS-01	Dimensioning.
		M-HVAC-NOTE-01	Notes, drawing title & #.
		A-BRDR-TITL	Sheet border/title block.
M-1.2	**Mechanical -- 2nd Floor HVAC**	A-WALL-PLAN-01	Walls & columns.
		A-EQUIP-01	Equipment & fixtures.
		M-HVAC-DUCT-01	Mech. room & air ducts.
		Etc.	

UNIFORM LAYERING SYSTEMS continued

A BASIC LAYER IDENTIFICATION SYSTEM continued

DWG #	SHEET TITLE	LAYERS	LAYER CONTENTS
M-2.1	**Mechanical -- 1st Floor Plumbing Supply**	A-WALL-PLAN-01	Walls & columns.
		A-EQUIP-01	Equipment & fixtures.
		M-PIPE-WATR-01	Water supply.
		M-PIPE-GASS-01	Gas supply.
		A-PIPE-NOTE-01	Notes, drawing title & #.
		A-BRDR-TITL	Sheet border/title block.
M-2.2	**Mechanical -- 2nd Floor Plumbing Supply**	A-WALL-PLAN-02	Walls & columns.
		A-EQUIP-02	Equipment & fixtures.
		M-PIPE-WATR-02	Water supply.
		Etc.	
M-2.1	**Mechanical -- 1st Floor Plumbing Drain**	A-WALL-PLAN-01	Walls & columns.
		A-EQUIP-01	Equipment & fixtures.
		M-PIPE-DRAN-01	Water drainage system.
		M-PIPE-SEWG-01	Sewage drain system.
		M-ROOF-DRAN-01	Roof drains.
		M-FLOR-DRAN-01	Floor drains.
		A-DRAN-NOTE-01	Notes, drawing title & #.
		A-BRDR-TITL	Sheet border/title block.
M-2.1	**Mechanical -- 2nd Floor Plumbing Drain**	A-WALL-PLAN-02	Walls & columns.
		A-EQUIP-02	Equipment & fixtures
		Etc.	
E-1.1	**Electrical -- 1st Floor Power**	A-WALL-PLAN-01	Walls & columns.
		A-DOOR-01	Doors.
		A-EQUIP-01	Equipment & fixtures.
		E-POWR-CIRC-01	Power circuits.
		E-POWR-01	Wall, floor & equip. outlets panel boxes, etc.
		E-POWR-NOTE-01	Notes, drawing title & #.
		A-BRDR-TITL	Sheet border/title block.

UNIFORM LAYERING SYSTEMS continued

A BASIC LAYER IDENTIFICATION SYSTEM continued

DWG #	SHEET TITLE	LAYERS	LAYER CONTENTS
E-1.2	**Electrical -- 2nd Floor Power**	A-WALL-PLAN-02	Walls & columns.
		A-DOOR-02	Doors.
		Etc.	
E-1.1	**Electrical -- 1st Floor Lighting**	A-WALL-PLAN-01	Walls & columns.
		A-CLNG-01	Ceiling components, soffits, beams, slopes, etc.
		A-LITE-01	Light fixtures.
		E-LITE-CIRC-01	Circuits & switching.
		E-LITE-NOTE-01	Notes, drawing title & #.
		A-BRDR-TITL	Sheet border/title block.
E-1.2	**Electrical -- 2nd Floor Lighting**	A-WALL-PLAN-02	Walls & columns.
		A-CLNG-02	Ceiling components, soffits, beams, slopes, etc.
		Etc.	

We departed from the AIA Layer Guidelines several times in this simplified example. It's OK to vary the code system on smaller jobs as long as your layer names are well understood by your consultants. Otherwise, you should adhere to the standards proposed by the AIA.

The reason we deviated in this example is that some layer subdivisions and names offered by the AIA are more elaborate than needed by smaller projects.

If the AIA Layer Guidelines were applied to small-scale residential work, for example, then plumbing layers would include P-DOMW-EQPM for Domestic Hot and Cold Water Equipment, P-DOMW-HPIP for Domestic Hot Water Piping, P-DOMW-CPIP for Domestic Cold Water Piping and D-DOMW-EQPM for Domestic Hot and Cold Water Equipment. It would be a bit much.

TITLE BLOCKS

The title block hasn't gotten much respect in years past. Its importance was often unappreciated, and its design was relegated to the most junior of junior drafters. It seems incredible in retrospect, but up until barely ten years ago, most title blocks were individually hand drawn.

Now the title block is recognized as a key index for the contractor. It contains important legal information. It provides drawing preparation and revision sequence information that may be important in later litigation. And it would be nice if contractors could expect to find information in predictable places when they look at a working drawing sheet.

TITLE BLOCK FORMAT

The recommended and most common title block format is a vertical right hand strip, 3" wide. (An adjacent 3" strip to the left is for extended title block information, if necessary, such as building department or client agency approvals and/or general notes or keynotes).

Some offices strongly prefer a vertical format -- all data reading to the right of the sheet except for the sheet number which is always to be read vertically.

A few offices still use a block at the right hand corner of each sheet. This can be designed to fit within one or two 6" x 5-3/4" sheet modules but for general readability and ease of use, this is the least preferable.

Here is the recommended sequence of information, from top to bottom:

__ Primary Design Office Logo.

__ Design Office(s) name(s), address, phone, fax, email.

__ Consultant Sheet Identification.

__ Space for license stamps: Primary Design Office
    Stamp or Registration Number; Consultant Stamp.

__ Project or Client Identification:
    Client or Project Logo.
    Building Name.
    Phase.
    Client.

__ Issue dates: Phase, Addendum, Clarification, Revision.

__ Management & Filing Data: Project or Job Number,
    CADD file number.

__ Personnel: Drafter(s), Supervisor or Checker.

__ Sheet Title: Site Plan, First Floor Plan, etc.

__ Sheet Number: According to the uniform system.

More detail on the content and size of each portion:

Design office name/logo.

Consultants Identification.

Space for registration stamps.

Project Name.

Client Name.

Release Dates.

Management Data.

Sheet Title.

Sheet Number.

DETAILED CONTENTS CHECKLIST: TITLE BLOCKS

__ PRIME DESIGN FIRM'S NAME, ADDRESS, TELEPHONE, FAX, E-MAIL NUMBERS; PROJECT WEB SITE ADDRESS.

__ ASSOCIATED OR JOINT VENTURE FIRM NAMES, ADDRESSES, TELEPHONE AND FAX NUMBERS

__ REGISTRATION NUMBERS OR OFFICIAL STAMPS OF PRIME AND ASSOCIATED FIRMS

__ PROJECT NAME AND ADDRESS

__ OWNER'S NAME AND ADDRESS

__ CONSULTANTS' REGISTRATION NUMBERS, NAMES, ADDRESSES, TELEPHONE, FAX, AND E-MAIL NUMBERS (Include on title block if list is small. Otherwise list all data on the General Information sheet.)

__ DRAWING TITLE AND SCALE

__ MINI KEY PLAN OF BUILDING

__ CONTACT PEOPLE IN THE PRIMARY DESIGN FIRM. (This data may be on the General Information Sheet instead of the title block.)

  __ Principal in charge
  __ Chief or project architect
  __ Job or team captain
  __ Office phone number and extension, plus alternate or after hours phone number of primary contact for client, bidders, and contractor

__ DESIGNERS AND DRAFTERS (A growing preference is to provide full names of job participants on the General Information Sheet, as well as the usual specific staff initials on individual sheets.)

__ INITIALS OR NAME OF DRAWING CHECKER

__ SPACE FOR REVISION DATES AND REVISION REFERENCE SYMBOLS

__ PROJECT NUMBER

__ FILE NUMBER

__ COPYRIGHT NOTICE OR NOTE ON RIGHTS AND RESTRICTIONS OF OWNERSHIP AND USE OF DRAWINGS

__ FINAL RELEASE DATE

DETAILED CONTENTS CHECKLIST continued

__ DRAWING SHEET NUMBER AND TOTAL NUMBER OF DRAWINGS

__ SHEET NUMBER CODING FOR CONSULTANTS' DRAWINGS

__ CODING FOR SEPARATE BUILDINGS OR PORTIONS OF PROJECT

__ BUILDING DEPARTMENT APPROVALS (Sometimes required on drawing sheets, sometimes only on the General Information Sheet.)

__ PERMIT NUMBERS

__ SPACE FOR APPROVAL STAMPS OR INITIALS

__ DATES OF APPROVAL

__ PRIVATE BUILDING CODE CHECKING SERVICE

__ CHECKING DATES

__ JOB PHASE COMPLETION DATES

__ CLIENT APPROVALS

__ BUILDING AND PLANNING PERMIT AUTHORITIES, ADDRESSES, TELEPHONE AND FAX NUMBERS

Notes

____

____

____

____

____

____

____

COVER SHEETS

The Cover Sheet is usually the very first sheet followed General Information Sheet. It provides an image of the project and a large display of the project name, client, prime design office and consultant.

There's no standard in content or format for these sheets, but they often feature a perspective drawing or photograph of a model, or in a few cases, photographs of the client team, the design and production team, or even the users of the building (such as a cluster of school children).

To whatever degree a format may be standardized, we suggest:

__ Project name and client name(s) or logo centered, large and bold. Project address. (Sometimes a location map is shown on the Cover Sheet, but this is best reserved for the General Information Sheet.

__ Prime design firm name and logo, with complete contact information, centered.

__ Consulting firms, spaced and centered below the prime design firm's name.

__ Date and place of issue.

GENERAL INFORMATION SHEETS

Like Title Blocks, the General Information Sheet also used to be a good "keep busy" assignment for junior drafters. Thus there has been little consistency in the type or quality of information provided, much less appearance. Oddly, the General Information Sheet, which provided clients and contractors with the first impression of overall drawing quality, often used to be the worst-looking sheet in the set.

Awkward hand lettering of sheet titles in the index; clumsy symbols and materials indication keys; abbreviations and nomen-clature too small or voluminous to ever be read; . . . that's all rapidly becoming history. General Information Sheets are now getting the attention they deserve.

A checklist of General Information Sheet content is provided on the next page.

As for format, since General Information Sheets vary in major ways from office to office and job to job, we can only suggest the most general of standards, such as the following:

__ The left-hand side includes the title block, with the sheet identification of G-1 or 00-1, depending on the system you select.

__ Top right: The module strip to the left of the title block is a convenient location the drawing sheet index. (This may be repeated in reduced size on subsequent drawings.) Include an explanation of the sheet numbering system, if it will help the contractor.

__ Middle right: Show permit and approvals information, if required. (Some required permit documentation may be in 8-1/2" x 11" format and will depart from the module grid.)

If a considerable amount of permit and approval documentation is required and it crowds the useful information, permit data can be provided on a second sheet reserved for such information: a sheet G-2 or 0.1 (Sometimes separate detailed ADA compliance or hazardous waste information may require even more General Information Sheets.)

__ Lower right: Building key plan, if any. A project location map with address, preceded if necessary by a regional vicinity map.

__ Middle third: General Notes, symbols legend, explanations of production systems and standards, such as explanation of dimensioning (whether plan dimensions are predominantly to framing or finish, unless noted otherwise).

__ Upper left-hand modules: production management data, layer schedule, file names, hourly work record.

__ Middle and lower left-hand modules: abbreviations, nomenclature.

Notes:

____

____

____

____

____

____

GENERAL INFORMATION SHEET

DETAILED CONTENTS CHECKLIST

__ PROJECT NAME

__ OWNER'S NAME

__ PRIME DESIGN FIRM

__ ASSOCIATED OR JOINT VENTURE DESIGN FIRMS

__ CONSULTANTS
Names, addresses, telephone, fax, and e-mail numbers. Include on Title Block if list is small. Otherwise list all consultants and related data on the General Information Sheet.

- __ structural
- __ HVAC
- __ plumbing
- __ electrical
- __ lighting design
- __ soil testing
- __ civil
- __ landscaping
- __ interiors
- __ acoustical
- __ auditorium
- __ kitchen/food service
- __ design management
- __ construction management
- __ energy
- __ solar
- __ fire safety
- __ barrier-free handicap access

__ INDEX OF ARCHITECTURAL AND CONSULTANTS' DRAWINGS

__ GENERAL NOTES
Some states and local building jurisdictions require certification of compliance with energy conservation or disability laws and regulations.

GENERAL INFORMATION SHEET continued

DETAILED CONTENTS CHECKLIST continued

__ LEXICON OF ABBREVIATIONS AND NOMENCLATURE
The CSI is preparing recommended standards for universal use.

__ LEGEND OF CONSTRUCTION MATERIALS INDICATIONS AND DRAWING CONVENTIONS
The CSI is preparing recommended standards for universal use.

__ LEGEND OF SITEWORK, STRUCTURAL, ELECTRICAL, PLUMBING, AND HVAC SYMBOLS AND CONVENTIONS
Usually shown on consultants' drawings.

__ LEGEND OF SYMBOLS:
- __ Elevation point
- __ Revisions
- __ Column or module grid key
- __ Building cross section or partial section key
- __ Wall section key
- __ Detail key
- __ Window number key to window schedule
- __ Door symbol key to schedule
- __ Room or area number key to finish schedule
- __ Stair key
- __ Equipment schedule key
- __ Existing work, work to be removed, new work

__ EXPLANATION OF PRODUCTION SYSTEMS
- __ Sheet numbering identificaiton
- __ Keynote system
- __ Use of fixture heights schedules instead of interior elevations
- __ Wall construction schedule key symbol, if wall materials indications aren't shown
- __ Subdued shadow print background printing

__ EXPLANATION OF LAYERING IDENTIFICATION SYSTEM

PRODUCTION AND LAYER MANAGEMENT DATA

As a set of drawings progresses, printing dates are necessary, to keep track of check-print and job phases.

Printing time and date should be near the sheet's lower left drawing module (or block)

Other management data to include:

Sheet file name

Defaults

Line weight (widths)

Printer command system

Layers

Production time.

Notes:

____

____

____

____

____

____

____

PREPLANNING WORKING DRAWINGS

MINI-MOCK-UP WORKING DRAWINGS

Some of the oldest firms in the US and overseas have used the simplest of tools for nearly 100 years they wouldn't dream of doing otherwise. So have some of the nation's most advanced CADD offices. Meanwhile, oddly enough, many others have never heard of it.

The tool is the old-time, mini mock-up working drawing set -- and these days the oldtimer is more useful than ever.

The mock-up set is usually drawn on 8-1/2" x 11" sheets -- one-quarter size rough sketches of site plan, floor plans, elevations, details, etc. Every sheet to be part of an upcoming working drawing set is included.

"Why bother?" says the "show-me" manager. "My people have done so many working drawings, they could do them blindfolded."

Blindfolded is the word. Doing a set of drawings without a mock-up guide is like driving in a strange city without a roadmap. No matter how many times people have done working drawings, they are still full of false starts, redundancies, errors, and blank spots that wouldn't be there if the job had been planned from the outset.

Like all good things, the mock-ups have multiple values, any one of which would justify their use:

__ They're invaluable for doing more precise project time and cost estimates.

__ They help show the client the total Scope of Work, and have been used successfully to raise fees during client negotiations.

__ They guide drafters when supervisors aren't around. They keep drafters busy, help them draw the right work at the right scales in the right places, AND help them avoid duplicating or contradicting what others are doing.

__ They help guide the checkers during check-print time.

__ They keep CADD production under control -- a situation where work is often forgotten or overdone, because it's hard to see the whole job at once.

They have so many values that NONE of the top production offices we know ever starts a set of drawings without a mini-mock-up and a contents checklist to go with it.

Here's how users make a good thing better:

__ They preprint their working drawing sheet formats "long-ways" on oversized sheets, such as 11" x 17". They show the outline of the standard drawing size on the upper left corner, and add a drawing content and to-do checklist on the right-hand side.

__ They save time making the mock-ups by using reduction copiers to recycle reduced-size prints of design development drawings as working drawing sheets.

__ They clarify their CADD layering or manual overlay systems by including sample mock-ups of base sheets and overlays, to show how they actually relate.

__ They include drafting grid backgrounds on mock-up sheets, to assist freehand sketching. Mock-up sheets often also include office standard drafting modules (such as the 5-3/4" x 6" detail module).

__ The best organized firms use mock-ups to show the production enhancements that are supposed to be used on a job. If paste-up and photodrawing are to be used, for example, they are shown in the mock-up. If a fixtures height schedule is used instead of endless interior elevations, they are specified. The mock-up can be a sort of specification of graphic standards, as well as a "working drawing" for the working drawings.

__ Some supervisors use a copy of the mock-up to keep track of the hours spent on each working drawing sheet. That helps refine the office time and cost database, so future jobs can be more accurately estimated.

Mock-ups are especially fast and easy to make if you take advantage of the office enlargement-reduction copier. Since some elements will repeat themselves on the mock-up, just as they will in the drawings, there's no need to resketch them over and over. Just use photocopier paste-ups.

Don't forget to distribute copies of the mock-ups as a guide to all drafting staff on a job. And send out new copies when there are major changes in the composition of the drawings. Some mock-ups are created for planning and then disappear into a file. They lose half their value by lack of use throughout the project.

All tools get abused, and the best ones get beaten to death. For example, when a drafter is assigned to do a mock-up and does it with all the excessive overdrawing typical in architectural drafting, this is a supervision problem -- not a failure of the system.

The pages that follow show illustrations of the appropriate level of completion of sketched mini-mock-up working drawings.

PREPLANNING WORKING DRAWINGS continued

A typical hand sketched set of mini-mockup working drawings.

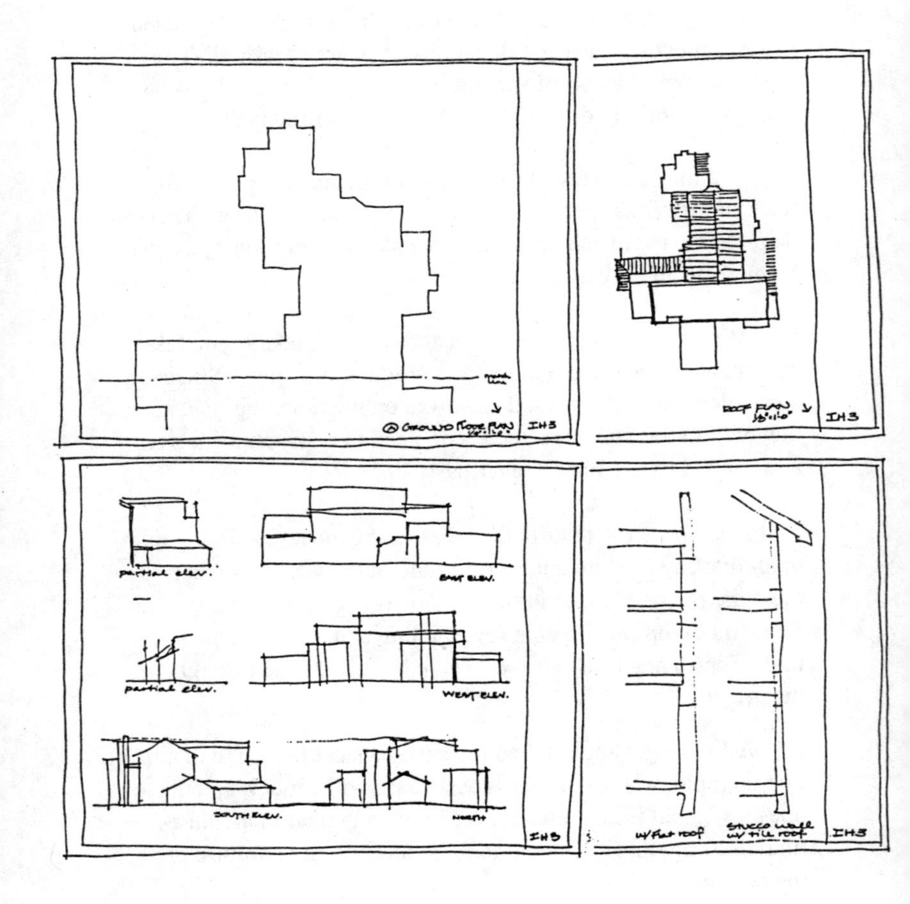

PREPLANNING WORKING DRAWINGS continued

Mockups are especially useful for organizing and planning detail sheets and wall sections.

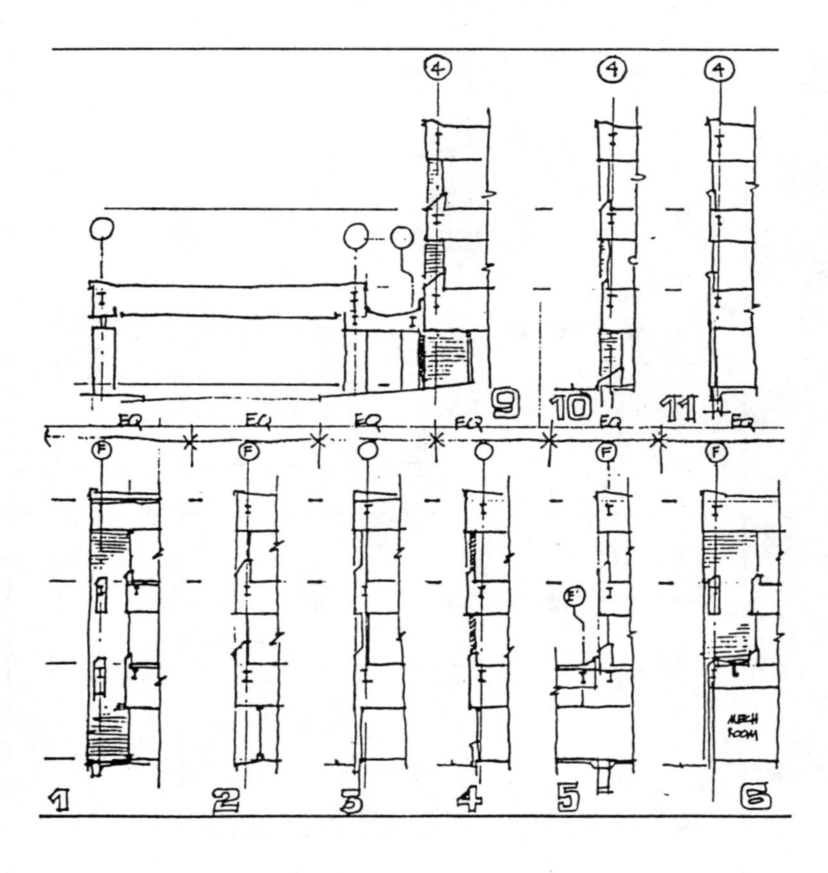

PREPLANNING WORKING DRAWINGS continued

Many firms create their mockup planning sheets on CADD and use them as standards for future projects. But most project managers still do hand drawn "cartoons."

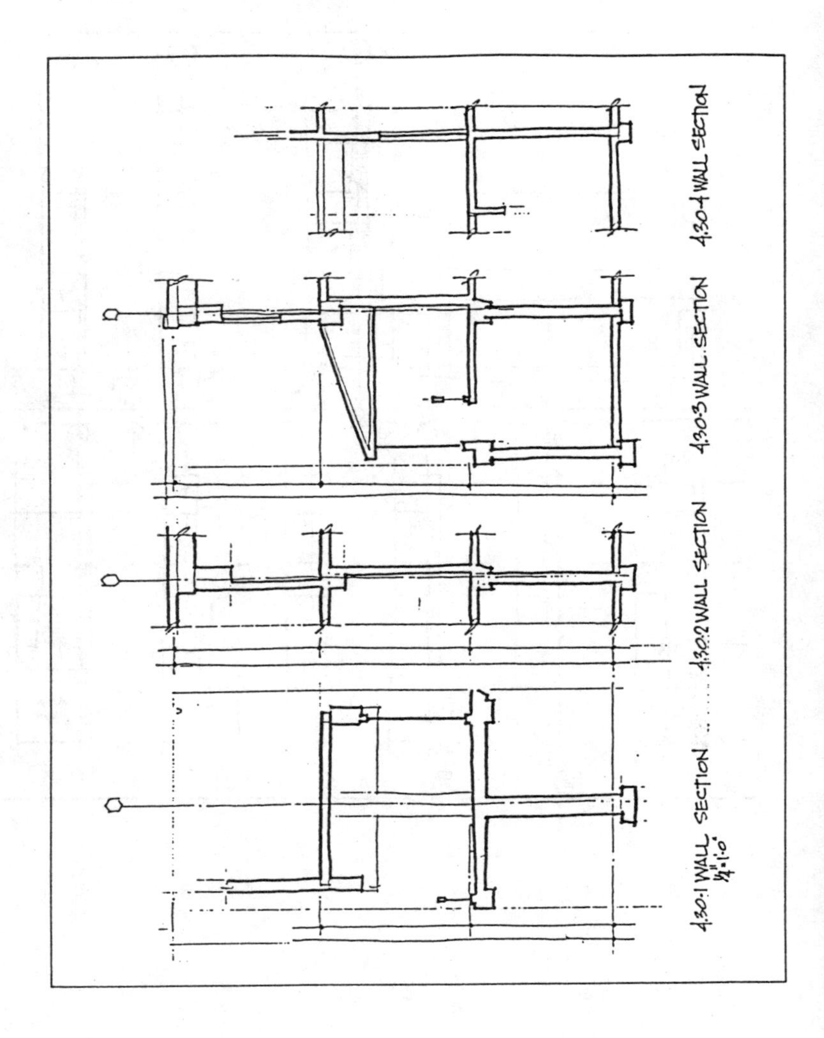

PREPLANNING WORKING DRAWINGS continued

A timesaver: Using reduced sized images of the presentation drawings for mockup plans, elevations and sections.

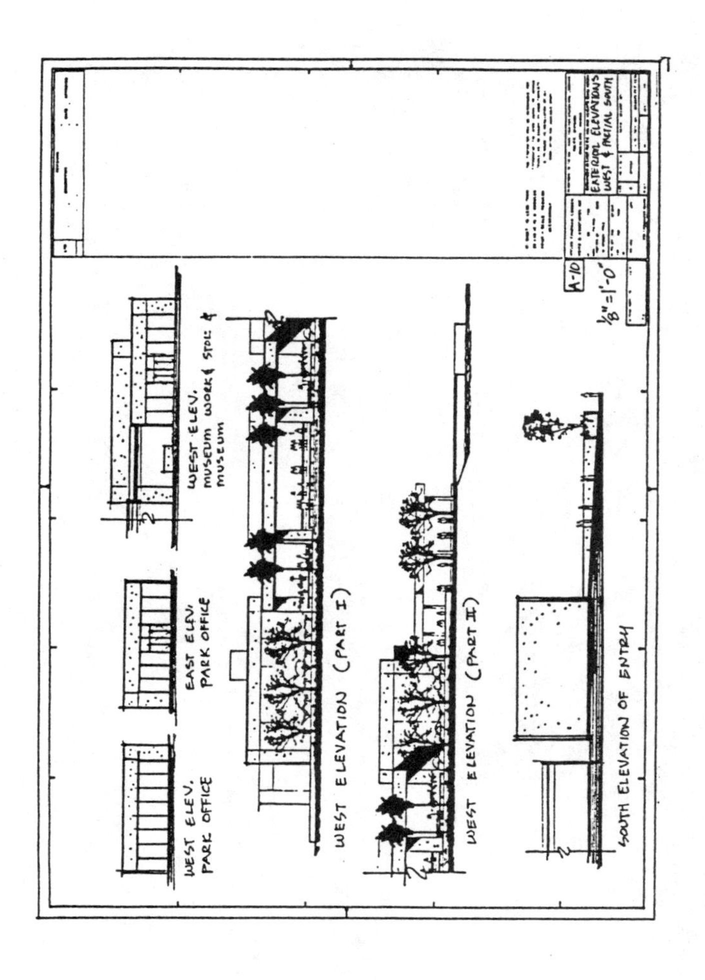

UNIFORM STANDARDS FOR NOTATION

FIVE KINDS OF ARCHITECTURAL NOTATION

Architectural notation is often too lengthy. Notes may describe materials in detail, product names, tolerances . . . information that should properly be in specifications. If the information is also in the specifications, it will either duplicate or contradict the notation.

The first rule in architectural notation is to keep it simple and generic. Unfortunately, because of the number of people who are likely to be adding notes to drawings, it doesn't always come out that way.

This is a good argument for using well-tested standard notation that drafters can just pick and choose from when doing their drawings.

While considering the possible advantages of a standard notation system in your office, consider the fact that working drawings include five different kinds of notes:

1) The easiest and most common notes are "IDENTIFICATION NOTES." They name in one, two, or a few words, a construction material, a component of construction, or a combination of materials and components.

Examples of IDENTIFICATION NOTES ARE: "Asphalt Paving," "Built-up Roofing," "Stucco," "1/2" Gypsum Wallboard." IDENTIFICATION NOTES are also sometimes called "TITLE NOTES."

All notation, whether keynotes, standard notes, or ad hoc traditional notes, should properly begin with their subject -- a material or object or product, or a combination, such as "Wood

Post." Usually the notes on large-scope drawings such as plans and elevations are of this simple type. Sometimes such notes are elaborated upon with dimensions, sizes, and spacings and/or referrals to other documents and "ASSEMBLY NOTES" (See item 4).

2) The second most common notes are "DIMEN-SION NOTES." These notes identify quantities, sizes, dimensions, locations, heights, and spacings of objects or materials named in the IDENTIFICATION NOTES.

Examples of "DIMENSION NOTES" are: "6 x 6 Wood Post," "8" Concrete Block," "4" Concrete Slab." Since there are infinite possibilities for DIMENSION NOTES, there is no need (nor any way) to standardize them.

Since numerical notes are short and simple in themselves, there is no reason to code them with any shorthand system. Just add them as required onto other notes or keynotes in the drawings.

3) The next most common note is the "See Note" or "REFERENCE NOTE." That's the note that says: "See Specifications," See Structural Drawings," See Detail 7/A22." These too are totally variable and are added to other notes as required on each drawing.

4) The fourth kind of note is the "ASSEMBLY AND FINISH NOTE." These notes follow an IDENTIFICA-TION or TITLE NOTE and explain in detail how components are arranged, fastened, treated, or finished. They are used mainly in construction details and sections and are not usually appropriate for small-scale, broad-scope drawings, such as floor plans or elevations. These notes make up the bulk of the multi-word Instructions notes on detail drawings and are also sometimes called "INSTRUCTION NOTES."

5) Another common kind of note states conditions or common instructions, like: "Existing, to be removed," "Slope to drain," "N.I.C." (Not In Contract). These are General Information notes, often added to Identification Notes. They are easily standardized but require a special coding method if you're using keynotes.

ASSEMBLY, FINISH, and INSTRUCTION NOTES are variable, depending on the specific construction task and your office's own design and construction standards. As you prepare your own office standard notation, you'll build a file of such notes, which you will code, file, and have ready for retrieval, editing, and reuse as needed.

ASSEMBLY NOTES AND MASTER NOTES

This section tells how to combine more elaborate construction information and/or instructions with your selected IDENTIFICATION NOTES. The more elaborate data are called ASSEMBLY NOTES, FINISH NOTES, or INSTRUCTION NOTES. We'll use the term ASSEMBLY NOTES and define them as: Notes that explain the arrangement, construction, connection, and/or finish of an item named in an identification note.

An example would be:

CONCRETE PAVING (02515)

Slope to drain, 1" in 10'

Or:

CONCRETE PAVING (02515)

2 x 6 redwood construction

joints @ 20' ea. way

PRELIMINARY STEPS FOR WRITING ASSEMBLY NOTES AND/OR MASTER NOTES

1) Every ASSEMBLY NOTE, whether consisting of one word or a dozen, must be preceded by its IDENTIFICATION or TITLE NOTE.

2) Each assembly's IDENTIFICATION NOTE should be followed by its CSI coordination number, as shown in the notes throughout this checklist. That automatically correlates the note with pertinent specification sections, standard details on file, and any other reference data you want to integrate through the CSI Master Format numbering system.

3) After or below the IDENTIFICATION NOTE, you may add a REFERENCE NOTE instructing the user to see another source for more information.

4) You may add size, location, or dimension note if your placing the note in or next to the field of the drawing. But if the drawing is keynoted, the note will stand alone in the Keynote Legend, and the dimension information will be added at the keynote reference number near the object being noted. That way one note can suffice for any number of times the object appears, regardless of differences in size, location, spacing, or dimension information.

5) A GENERAL INFORMATION NOTE, such as "Existing, to be repaired," may follow an IDENTIFICATION NOTE.

6) If you are creating a file of standard or Master Notes, you will add an extra number after each note's CSI Coordination Number. That's required, because there may be more than one ASSEMBLY NOTE, and it will have its own coding for filing and retrieval purposes. From the above examples, the

numbers .1 and .3 have been added to permanently identify and separate ASSEMBLY NOTES from one another.

CONCRETE PAVING (02515.1)

Slope to drain, 1" in 10'

CONCRETE PAVING (02515.3)

2 x 6 redwood construction joints @ 20' ea. way

Henceforth, although the CSI coordination number 02515 will always mean CONCRETE PAVING, only the number 01515.1 will mean that particular drainage slope with such paving.

The numbers ".1" and ".3" are used here for illustration; your suffix numbers would likely be quite different. And please keep in mind that these are CSI Coordination Numbers and FILE numbers for finding stored notes, NOT Keynote numbers.

WRITING ASSEMBLY NOTES

After an IDENTIFICATION NOTE and related data are in place, you may need to elaborate on the information by telling what is to be done with the object in terms of assembly or finish.

Although ASSEMBLY NOTES are used mainly in detail drawings, architects increasingly favor giving the contractor as much data as possible throughout all the drawings. Advanced keynoting and computerized Master Notation make it possible to provide more thorough ASSEMBLY NOTES than ever be fore, at no added time or cost, and without losing control of drawing and specification coordination. Here's The Golden Rule of notation writing, followed by some explanations:

EVERY NOTE SENTENCE HAS EITHER A SUBJECT ONLY, OR A SUBJECT PLUS A PREDICATE. That is, there is something you have named -- the subject -- and that subject may be followed by something you say about it. For clarity, keep each note sentence restricted to one subject and one general predicate.

This rule is very useful for sorting out what should and should not be written in assembly notation.

What is usually said about any subject is that something else has a relationship to it. That relationship may be one of action, as in the note that "(the contractor shall) Patch existing plaster to match texture of adjacent walls." Or the relationship may be in terms of physical proximity or connection to other things, such as "(the contractor shall attach/assemble) 3/16" resawn redwood plywood over 1/2" gypsum board w/ 3/8" x 3" redwood battens."

Observe that this latter note is not really an ASSEMBLY or FINISH NOTE. It's actually a list of IDENTIFICATION NOTES strung together as a sentence. And that is an example of what usually should not be presented as an ASSEMBLY NOTE.

The reason: IDENTIFICATION NOTES are limited in number and have been sorted out and properly classified for easy keynoting and standardization in this checklist. The numbers of possible note combinations, however, run to the millions. The only way to control the numbers and keep them manageable is to stick to a consistent, one-at-a-time, individual IDENTIFICATION NOTE system.

CONTENT AND DIVISIONS OF ASSEMBLY AND FINISH NOTATION

You start with the subject -- the IDENTIFICATION NOTE of a material or object -- and then name some action and/or related material (s) or object (s). Here are some examples:

CONCRETE PAVING (02515)
See Det. A1/9 Floated and steel
troweled finish

4 x 8 x 16 REINF. CONC. BLOCK (04230)
Rake horiz. joints

METAL HANDRAIL (05520) See Det. A1/5
2" diam. pipe rail
2" posts @ 6' o.c.
Cast-in-place pipe sleeve @ base (02444)

Connect handrail post bracket @ conc. wall
w/ 4-6" exp. bolts (03250)

In the first example, CONCRETE PAVING, the action is described in past tense, as if looking at the final result rather than telling the contractor what action to take. That's acceptable, although it's more readable and concise to refer to things in the active tense, like: "Float and broom finish." The thickness of the concrete slab might be included with the note and/or left for the detail.

The third example includes a secondary list of materials, which could have been treated as separate IDENTIFICATION NOTES. The only ASSEMBLY/FINISH NOTE involved is the last one about connecting the handrail post bracket. And even that could have been written as a separate IDENTIFICATION NOTE without the explicit instruction.

As you examine traditional working drawing notation, you'll discover that a large amount of what appears to be assembly instruction is really just a series of identifications of materials or objects. In fact when notation goes much beyond just naming what things are; it starts to duplicate or contradict the specifications. Thus the second Golden Rule:

USE SHORT, SEPARATE IDENTIFICATIONS OF MATERIALS AND PARTS IN SINGLE NOTES, IN PREFERENCE TO LONG COMBINATIONS OR LISTS OF MATERIALS AND PARTS.

A checklist is preferable for clarity. Now some Q & A:

Q: What if you're going through the checklist, marking notes on a checkprint, and you think of a note that isn't on the list. Where do you put the note?

A: Write the note out on the blank backside of the adjacent checklist sheet. As you find some notes that you especially favor but that are not in the checklist, insert them in the original master copy as a permanent part of your system.

Q: What if you think of a special IDENTIFICATION NOTE that is not in the particular section of the notation checklist you're working with? How do you find it and find its CSI Coordination Number without thumbing all through the checklists?

A: Jot down the note. Don't stop to look up the number; just proceed with the checklisting process. Later you will come across the special item in another portion of the list and can backtrack to fill it in. As a reminder of unfinished items, apply a yellow "post-it" flag or 3 x 5 card, wherever you have an uncompleted note in your list.

Q: When some part of a note isn't fully established yet, should you stop to work it out, or proceed with the checklist process?

A: Proceed without interruption. Mark the unknown or missing item as previously described. You'll move faster by sticking to one aspect of the process at a time. Missing or unknown items will most often resolve themselves as you continue checklisting.

SETTING UP A STANDARD NOTATION FILE

Standard notes serve the same purpose as standard details. There's no point to having staff members rewrite the same basic notes over and over. Instead, most notes can be created once, filed and then edited and reused for later projects.

Besides saving time and money, standardization enhances quality control. If you're selecting and creating notation for reuse, you can afford to give the notes a little extra time and attention. And, with feedback, you can upgrade and elaborate on them on the basis of accumulated jobsite and office experience.

How do you "file" a note for reuse?

Start with this checklist as your guide. You can insert wherever you please any single item IDENTIFICATION NOTE you want to use that is not in this checklist. We suggest you create extra pages, and, following the same format as these checklist pages, write in new IDENTIFICATION and ASSEMBLY/FINISH notation as you see fit and add new pages to the checklist as you see fit.

A note that is often included on site plans:

"Slope all exterior steps and walkways to drain."

That kind of note can have a GENERAL INFORMATION prefix keynote number of your choosing.

You may elaborate on a concrete paving note by saying something like:

CONCRETE PAVING (02515)

Construction joints at 8' each way, @ columns, and @ abutting curbs or wall construction.

If you frequently use that note on concrete paving, then make it a permanent part of your notation file in your Master Note binder.

What if you come up with a new IDENTIFICATION NOTE and don't know how to number it or where to file it?

As an example, you might do a lot of wood frame residential construction, always use leveling grout on foundation walls under 2 x 6 redwood mud sills, and anchor the sills with 1/2" diameter anchor bolts 10" long, @ 6' o.c. Strictly speaking, you could identify the concrete foundation wall as (03306), the redwood sill with a wood framing number, grout with another number, and the anchor bolts with still another number. But if you repeat the note is repeated on every job of this type, it's more economical to combine all these IDENTIFICATION NOTES as a single, standard assembly note. Then, getting back to the question above, how do you identify and number it? You have to decide if it's a concrete note, framing note, or anchor bolt note.

You can set your own rules, but we suggest you identify and number an original standard note by the way it will be titled. For example, the above note would have as its Identification, CONCRETE FOUNDATION WALL, because that's the subject. That's the subject because the other items -- mudsill, grout, and anchor bolts -- are there secondarily to, and added to, the foundation. The foundation wall can then be treated as the PRIMARY, all other related notes as secondary.

You establish priorities, primaries, and secondaries, by identifying the functional relationships of "what supports what." That serves as your guide for identifying the kind of note you're dealing with in any particular instance. That, in turn, tells you how to number new, large ASSEMBLY NOTES you might create.

GENERAL INFORMATION NOTES

GENERAL INFORMATION NOTES are those often-repeated general descriptions and instructions that don't relate to materials or construction products.

Examples are:

"Supplied by Owner, Installed by Contractor."

"Unless noted otherwise."

"Dimensions shown are approximate."

Many such notes are efficiently identified with commonly used abbreviations.

For example:

"Unless noted otherwise" is usually "U.N.O."

"Not in contract" is "N.I.C."

"On center" is "o.c."

It's an open question whether these easily abbreviated items should be spelled out in keynote legends. They are usually appended to other notes, so it makes sense to just add the abbreviations to a keynote or keynote reference number.

Notes:

KEYNOTING

Keynoting is the use of numbers in a drawing which refer to complete notation provided in a keynote list at the side of a drawing sheet.

Here is an example.

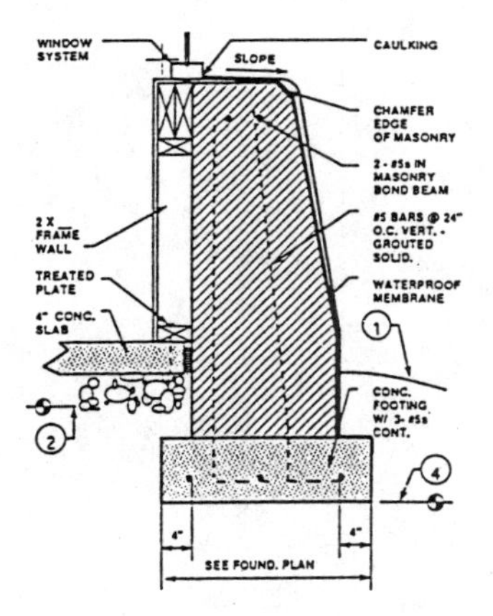

NOTE KEY

1 FINISH GRADE.

2 COMPACTED ROUGH GRADE.

3 12" MIN. BELOW UNDISTURBED SOIL OR ENGINEER CERTIFIED COMPACTED SOIL.

4 18" MIN. BELOW UNDISTURBED SOIL OR ENGINEER CERTIFIED COMPACTED SOIL.

Since many drawings use the same notes over and over again, it make sense to exploit this repetition through standardization.

The problem with keynotes is they tend to be misused and used where traditional notation would do the job better.

Misuse was evident in the ConDoc system promulgated several years ago where keynote numbers, instead of being simple one or two digit numbers, were full five digit's +, sometimes as long as 8 numbers.

There's no way a contractor or worker in the field could to find an 8 digit keynote number in the field of a drawing and remember it long enough track it to a separate keynote list.

BASIC KEYNOTING

The simplest keynotes: Plain 1, 2, 3, . . . numbers linking drawing to notation legend.

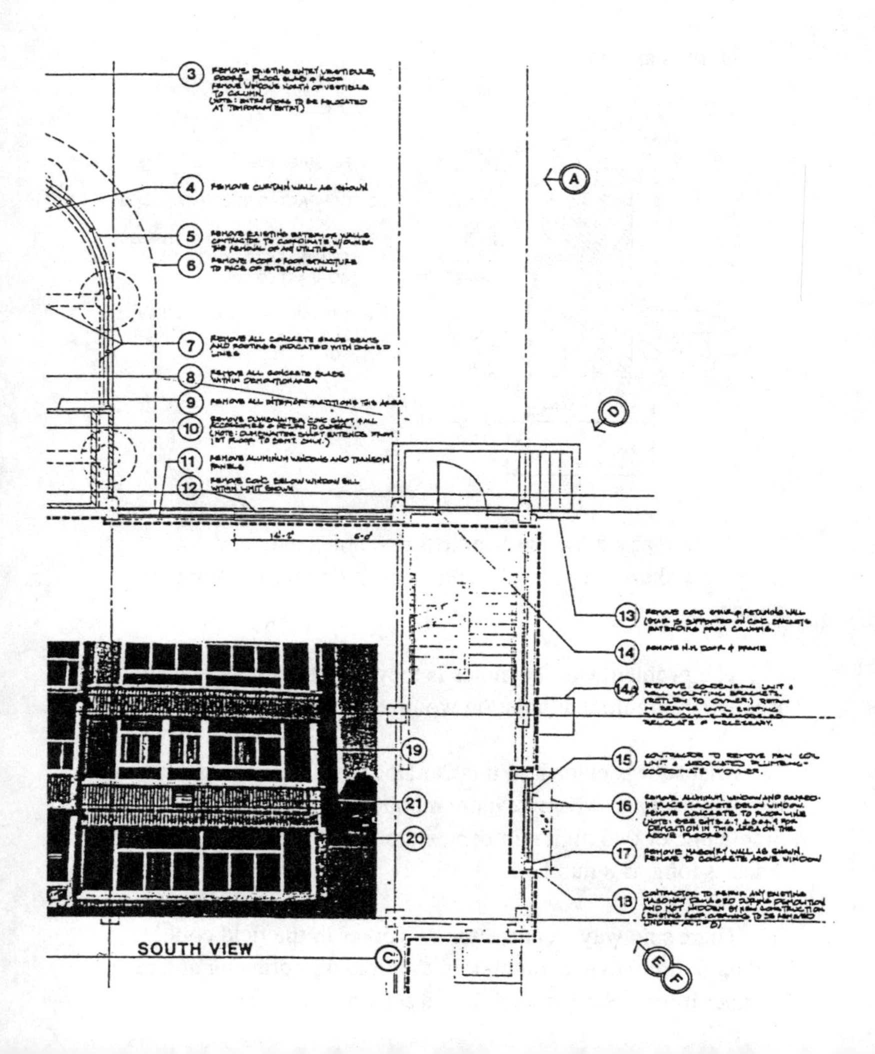

SYSTEMS NOTES

An illustration of standard "systems" notes -- notes that are common to most projects. They're consolidated as standards that are edited and then pasted into working drawings.

• 6" RATED PARTITION
SAME AS WALL TYPE "W-4" w/ 2" x 6" STUDS.

• WALL BETWEEN POOL AREA & LOBBY:
1/2" G.W.B. (lobby side)
FRAMING AS NOTED
R-12 BATT INSULATION (NO VAPOUR BARRIER)
1/2" "DENS-SHIELD" BACKER BOARD

• WALL BETWEEN ELEV. LOBBY & POOL AREA
5/8" TYPE "X" GYPSUM BOARD, JOINTS TAPED & FILLED
RESILIENT CHANNELS @ 24" o/c
2" x 6" STUDS @ 16" o/c
R-12 BATT INSULATION
5/8" TYPE "X" DENS-SHIELD (NO VAPOUR BARRIER)
1 HOUR SEPARATION: Sim. to U.L.C. #W301

• 8" EXTERIOR WALL @ TOWER
STUCCO FINISH ON MESH
BUILDING PAPER
1/2" EXTERIOR GRADE GYPROC
1 3/8" x 5 5/8" METAL STUDS @ 16" o/c BETWEEN H.S.S. COLUMNS

• WALL @ FITNESS ROOM/POOL EQUIP. ROOM
1/2" DENS-SHELD
2" x 4" FRAMING @ 16" o/c
1/2" DENS-SHELD

KEYNOTES CLUSTERED BY CSI DIVISION

An improvement on the concept: Keynotes remain simple -- 1 to 4 digits -- but they are preceded with a prefix number that identifies the relevant CSI division. Thus sitework notes start with a 2, concrete notes with a 3, etc.

KEYNOTES

1. GENERAL

1.9 TYPICAL FORMED PARAPET:
WOOD FRAMING W/GENERAL
CURVED SHAPE FORMED WITH
EXPANDED METAL LATH AND
STUCCO SYSTEM.

1.10 TYPICAL FREE STANDING EXTERIOR WALL:
CONC. FOOTING
C.M.U. WALL
WATERPROOF MEMBRANE BELOW GRADE
STUCCO SYSTEM (1/2" THICK MIN.)

4. MASONRY

4.1 C.M.U. WALL
4.2 C.M.U. COLUMN
4.3 C.M.U. FIREPLACE MASS
4.4 GLASS BLOCK
4.5 CUT STONE
4.6 STONE VENEER

5. METALS

5.1 STEEL RAILING
5.2 HANDRAIL
5.3 METAL DECKING
5.4 METAL FRAMING

COMPREHENSIVE BUT STILL SIMPLE

Suited to larger projects, these keynotes are simple, clustered according to CSI category, and they include a complete CSI division and section number in parenthesis after the note. This system greatly enhances drawing-specifications coordination.

KEY NO.		KEYNOTE
7.26		STANDING SEAM MET ROOF - 07410
7.32		CLOSURE/DRIP FLASHING - 07410
7.33		MET SOFFIT, COLOR TO MATCH ROOF - 07410
7.34		REUSE EXIST RETAINING BAR & FLASHING. RE-SET BAR IF REQ'D 07410
	**	AL FLASHING & TRIM - 07600
7.36		AL SILL FLASHING - 07600
	**	SEALANTS - 07900
7.50		SILICONE CAULK W/BACKUP - 07900
7.58		SET THRESHOLD IN BED OF SEALANT 07900
7.70		CAULK & BACKUP, BOTH SIDES - 07900
7.76		RE-CAULK RETAINING BAR & FLASHING - 07900
	**	MET DRS & FRAMES - 08110
8.10		HOLLOW MET DR - 08110
8.16		HC MET TRANSOM PANEL - 08110

WHAT NOT TO DO

Avoid using large CSI numbers as the keynote reference numbers. They're hard to use and some building departments will not accept drawings with this type of keynoting. (Neither will some contractors, unless they sense they'll get some change orders and extras out of the deal.)

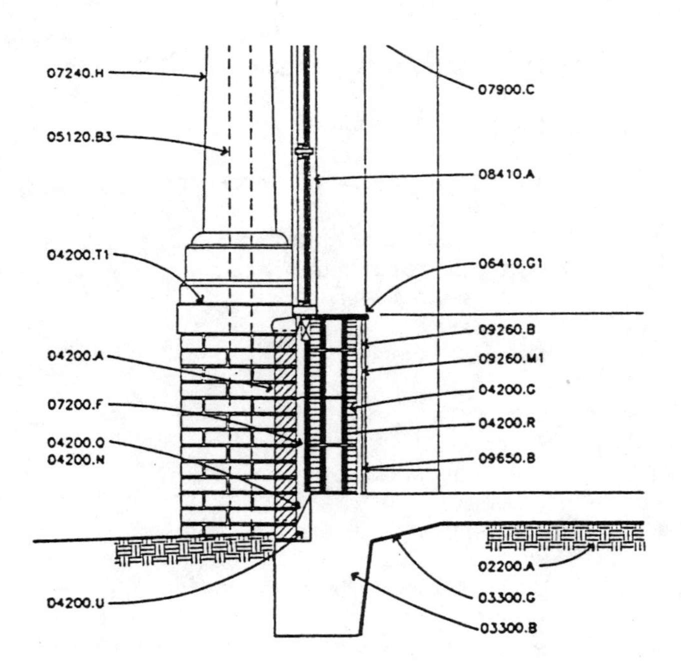

UNIFORM DIMENSIONING FORMAT

Floor plan dimensioning should be from framing to framing in most cases. Dimensions that are from finishes to finishes have to be recalculated (sometimes in error) by the framer. If a finish dimension is needed too, it can be

Ceiling heights should be designated as from finish floor or sub floor. It's usually to finish ceiling. Schedules with room names and ceiling heights such as the ones below are common.

Cross section and exterior elevation dimensions are normally from framing to framing or substructure unless designated otherwise.

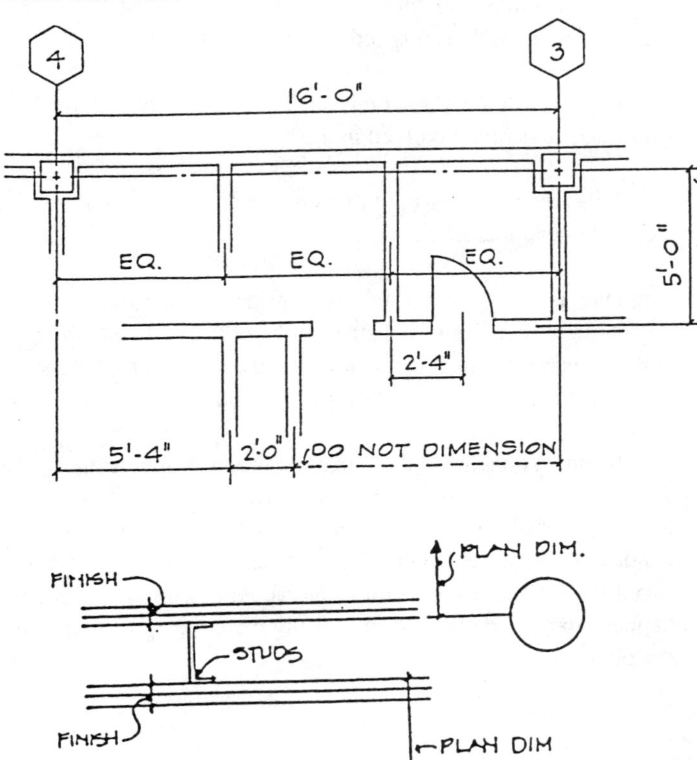

Other recommendations regarding dimensioning are listed below. These are from the Northern California Production Office Procedures manual which, while many years old, remains quite current.

FROM THE P.OP. MANUAL

Dimensioning should start with critical dimensions as predicated by design or other requirement.

It should take into consideration the trades using them and the sequencing to their respective work.

Keep in mind that tolerances in actual construction can be as varied as the people involved in the construction process.

This means that as-built dimensions do not always coincide with design dimensions.

Dimensioning from established grids or structural elements, such as columns and structural walls, assists the trades that, for example, must locate some of their work prior to the placement of floor slabs and of the partition layout that follows.

It can be concluded then, that the practice of closing a string of dimensions running between two reference points, might be useful for office checking, but it can definitely be conducive to conflict when, as in another example, the work of two different trades which must interact at a common point happens to be laid out by each starting from the opposite end of the dimensional string.

UNIFORM DIMENSIONING FORMAT continued

1) All numbers 1/8" high.

2) Fractions to have diagonal dividing line between numerator and denominator.

3) Dimensions under 1'-0" shall be noted in inches, i.e., 11", 6", etc. Dimensions 1'-0" and over shall be expressed in feet.

4) Fractions under one (1) inch shall NOT be preceded by a zero.

5) Dimension points to be noted with a short blunt 45° line. Dash to be oriented the same for vertical and horizontal runs of dimensions. Modular dimension points may be designated with an arrow or a dot.

6) Limit fractional dimensions in plans and elevations to 1/8" except for indication of a single material thickness.

7) Dimension all items from an established grid or reference point and do not close the string of dimensions to the next grid or reference point. Dimensioning shall be started with critical dimensions as predicated by design or other requirements. Since there is always the possibility of a variance between the on-job conditions and the design dimensions, all trades on the job should lay out their respective work from the same reference point. (Check non-dimensioned spaces for adequacy.) If three equal spaces are required in a given 12'-0" space, note it as such rather than noting three 4' dimensions.

8) Dimension; to face of concrete or masonry work; to centerlines of columns or other grid points and to centerlines of partitions. In non-modular wood construction dimension to critical face of studs. When a clear dimension is required either by code (or other reason), dimension to the finish faces and note as such. Do not use word "clear". Likewise, furred spaces

should be dimensioned from face to face or from a structural point to finish face.

9) Dimension as much as possible from structural elements rather than from items that may not yet be installed when the layout takes place, i.e., for plumber or electrician laying out sleeves on the forms for the floor deck.

10) Do not dimension items such as partitions or doors, that are centered or otherwise located on a grid, module, mullion, by schedule or by typical detail condition. The general notes or typical details should cover this fact or any other typical condition of dimension. Dimension unscheduled openings in accordance with paragraph 8 above.

Notes:

SCHEDULES

Schedules -- finish schedules, door schedules, window and equipment schedules -- are potentially great time savers but they often become so complicated that their original purpose gets lost. In this chapter we'll recommend some schedule formats as superior to others and urge that they be considered as part of a nationally accepted uniform drawing format.

Below is a good, simple standard door schedule to compare with others shown on the folowing pages.

DOOR SCHEDULE				
MK	SIZE	TYPE	MATERIAL	REMARKS
1	PR 3'-0 x 7'-0"	FULL LITE	ALUM. W/ INS. GLASS	NOTE 1, 4, 5
2	3'-0 x 7'-0" x 1 3/4"	FLUSH	18 GA. STEEL	NOTE 2
3	3'-0 x 7'-0" x 1 3/4"	FLUSH	18 GA. STEEL	NOTE 2
4	3'-0 x 7'-0" x 1 3/4"	FLUSH	18 GA. STEEL	NOTE 2
5	3'-0 x 7'-0" x 1 3/4"	FLUSH	18 GA. STEEL	NOTE 2, 3
6	3'-0 x 7'-0" x 1 3/4"	FLUSH	18 GA. STEEL	NOTE 2
7	3'-0 x 7'-0" x 1 3/4"	FLUSH	18 GA. STEEL	NOTE 2
8	3'-0 x 7'-0" x 1 3/4"	FULL LITE	ALUM. W/ INS. GLASS	NOTE 1, 4, 5
9	3'-0 x 7'-0" x 1 3/4"	FLUSH	18 GA. STEEL	NOTE 2
10	3'-0 x 7'-0" x 1 3/4"	FLUSH	18 GA. STEEL	NOTE 2

Below is a partial example of a commonly used complex finish schedule format. Sometimes this format requires several full size sheets in a set of drawings.

ROOM FINISH SCHEDULE

NO	ROOM NAME	FLOOR		WALLS								CEILING		NOTES
		MTL	BASE	NORTH	C	EAST	C	SOUTH	C	WEST	C	MTL	HEIGHT	
106	Commons	CPT1	VB	EXG PLAS	P8	--	--	--	--	GPDW/GL	P1	ACT	--	3,11
107	Abstract	CPT1	VB	GPDW	P1	EXG PLAS	P1	GPDW	P1	EXG PLAS	P6	ACT	--	3
108	Vestibule	EXG CT	EXG	EXG		EXG		EXG		EXG		EXG	EXG	
109	Womens Restroom	EXG CT	EXG	EXG	--	EXG	--	EXG	--	EXG	--	EXG	EXG	
110	Womens Restroom	EXG	EXG	EXG	--	EXG	--	EXG	--	EXG	--	EXG	EXG	
111	Mens Restroom	EXG	EXG	EXG	--	EXG	--	EXG	--	EXG	--	EXG		
112	Corridor	CPT1	VB	EXG		EXG PLAS	P1	--	--	EXG PLAS	P8	EXG PLAS	--	11
113	Janitors Closet	EXG T2	EXG	EXG		EXG		EXG		EXG		EXG	--	
114	Mens RR Vestibule	EXG CT	EXG	EXG	--	EXG	--	EXG	--	EXG	--	EXG PLAS	EXG	
115	Mens Restroom	EXG CT	EXG	EXG CT	--	EXG CT	--	EXG CT	--	EXG CT	--	EXG PLAS	EXG	
116	Corridor	CPT1	VB	--	--	EXG PLAS	P1	GPDW	P11	GPDW	P1	ACT	8'0"	3
117	Cataloging	CPT1	VB	GPDW	P7	GPDW	P1	GPDW	P1	EXG PLAS	P1	ACT		3
118	Catalog Office	CPT1	VB	EXG PLAS	P7	GPDW	P1	GPDW	P1	EXG PLAS	P1	ACT		3,2
119	Aquisitions	CPT1	VB	EXG PLAS	P7	GPDW	P13	GPDW	P1	GPDW	P1	ACT		3
120	Aquisitions Office	CPT1	VB	EXG PLAS	P7	GPDW	P1	GPDW	P1	GPDW	P1	ACT		3
121	Corridor	CPT1	VB	GPDW	P1	GPDW	P1	EXG PLAS	P6	GPDW	P1	ACT	8'0"	3
122	Reference Collection	CPT1	VB	EXG PLAS	P8	GPDW	P1	--	--	GPDW	P1	ACT		3,11
123	Public Catalog	CPT1	VB	--	--	--	--	--		EXG PLAS	P8	ACT		11
124	Microform	CPT1	VB	GPDW	P1	GPDW	P1	GPDW	P1	GPDW	P1	ACT	8'0"	
125	Office	CPT2	VB	GPDW	VWC1	GPDW	VWC1	GPDW	P1	GPDW	P1	ACT	8'0"	
126	Conference	CPT2	VB	GPDW	VWC3	GPDW	VWC	GPDW	P1	GPDW	P1	ACT	8'6"	
127	Lounge	CPT1	VB	GPDW	P1	GPDW	P6	GPDW	P1	GPDW	P1	ACT		
128	Copiers	CPT1	VB	GPDW	P1	GPDW	P2	--		--	--	ACT		
129	Secretary	CPT2	VB	GPDW	P1	GPDW	VWC1	GPDW	P1	GPDW	P1	ACT	8'0"	
129A	Storage	VAT	VB	GPDW	P2	GPDW	P2	GPDW	P2	GPDW	P2	ACT	8'0"	PLAM #1
130	Office	CPT2	VB	GPDW	P1	GPDW	VWC2	GPDW	P1	GPDW	P1	ACT	8'0"	
131	Corridor	CPT1	VB	GPDW	P6	GPDW/GL	P1	GPDW	P1	GPDW	P1	ACT	8'0"	
132	Corridor	CPT1	VB	GPDW	P1	--	--	GPDW	P1	--	--	ACT	8'0"	
132A	Storage	VAT	VB	GPDW	P1	GPDW	P1	GPDW	P1	GPDW	P1	GPDW	8'0"	
133	Check Out	CPT1	VB	GPDW	P1	--	--	GPDW	P2	--	--	ACT	8'0"	

THE RECOMMENDED FINISH SCHEDULE

In this system there's a code letter below the room name (D under OFFICE, C under RECEPTION). The code letter on the finish schedule represents the entire range of finishes in the room -- floor, base, walls, ceiling.

This simple consolidation creates a pocket size schedule that can convey the same information as a traditional full-sheet size finish schedule.

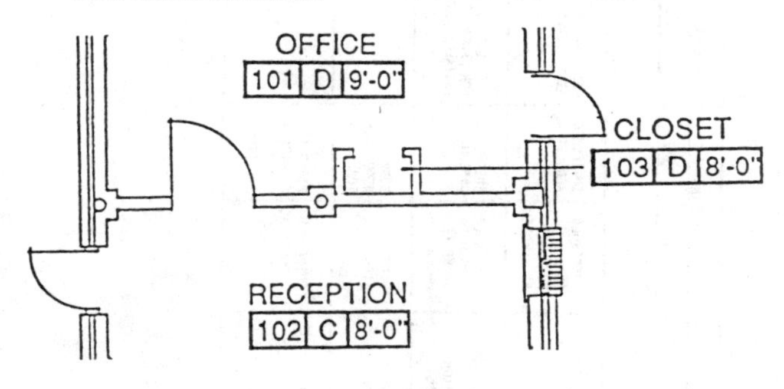

FINISH SCHEDULE					
FINISH KEY	FLOOR	BASE	WALLS	CEILING	REMARKS
A	CONCRETE	NONE	CMU	NONE	ROOM 102 NOT TO BE PAINTED
B	CONCRETE W/HARDNER	VINYL	CMU	24" X 24" ACOUSTICALBOARD	OMIT HARDNER IN ROOM107
C	CARPET	VINYL	GYPSUM BOARD	ACOUSTICAL TILE	
D	CARPET	VINYL	GYPSUM BOARD	GYPSUM BOARD	
E	VINYL	VINYL	GYPSUM BOARD CMU	GYPSUM BOARD	
F	TERRAZZO	TERRAZZO	GYPSUM BOARD	GYPSUM BOARD	
G	ACCESS FLOORING	VINYL	WOOD PANELING	GYPSUM PLASTER	SEALER ON CONC. BELOW ACCESS FL.
H	BRICK PAVERS	WOOD	WOOD PANELING	ACOUSTICAL PLASTER	
J	CERAMIC TILE	CERAMIC TILE	VINYL WALL COVERING	24" X 48" ACOUSTICAL BOARD	
K	QUARRY TILE	QUARRY TILE	CERAMIC TILE	ACOUSTICAL TILE	

EXPANDED ROOM FINISH SCHEDULE FORMAT

People have argued that more complex work, such as rehab and remodeling, require a complex finish schedule. Below is the basic system we recommend, with extended format to allow for Existing and New work. The schedule still remains very simple and small enough to be included on the floor plan sheets.

FINISH SCHEDULE

FINISH KEY	FLOOR		BASE		WALLS		CEILING		REMARKS
	EXISTING	NEW	EXISTING	NEW	EXISTING	NEW	EXISTING	NEW	
A	EXISTING FINISHES TO REMAIN - NO WORK REQUIRED IN THESE AREAS								
B	VINYL ASBESTOS	EXISTING TO REMAIN	VINYL	EXISTING TO REMAIN	GYPSUM BOARD	EXISTING TO REMAIN	SUSPENDED GYPSUM BOARD TO BE REMOVED	SUSPENDED ACOUSTICAL	PATCH WALLS AS REQUIRED
C	VINYL ASBESTOS TO BE REMOVED	VINYL ASBESTOS	VINYL TO BE RE-MOVED	VINYL	GYPSUM BOARD TO BE RE-MOVED	GYPSUM BOARD	SUSPENDED GYPSUM BD. TO BE RE-MOVED	SUSPENDED GYPSUM BOARD	SEE (09650) FOR FLOORING INSTALLATION
D	CONCRETE	VINYL	NONE	VINYL	CONCRETE MASONRY UNITS	GYPSUM BOARD	EXPOSED STRUCTURE	SUSPENDED ACOUSTICAL BOARD	
E	TERRAZZO	PATCH AS REQUIRED-MATCH EXISTING	TERRAZZO	PATCH AS REQUIRED-MATCH EXISTING	VINYL WALL COVERING	EXISTING TO REMAIN	SUSPENDED ACOUSTICAL BOARD	PATCH AS REQUIRED-MATCH EXISTING	

ALTERNATIVE DOOR SCHEDULE FORMAT

Below is an alternative door schedule also originally proposed by the AIA P.O.P. manual.

DOOR SCHEDULE								
DOOR MARK	OPENING SIZE	TYPE (NOTE 2)	THICKNESS (3)	CONSTRUCTION (4)	FACING & FINISH (5)	GLASS (6)	RATING (7)	FRAME (8)
1	3'-6" x 7'-0"	A	✓	✓	✓	CW	1½	26

1. "✓" SHOWN ON SCHEDULE INDICATES TYPICAL
2. DOOR TYPES:

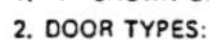

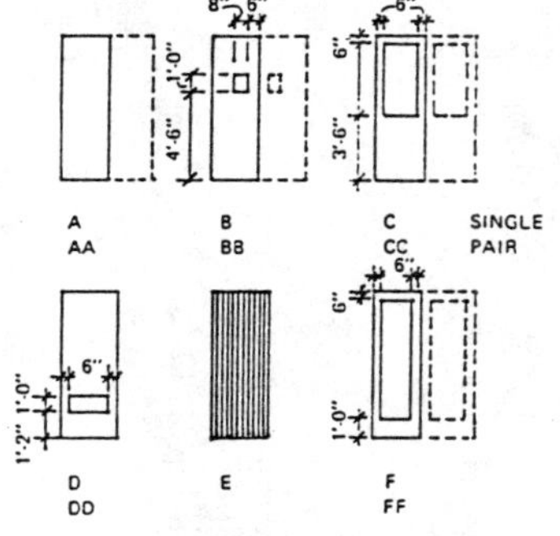

DOOR SYMBOL

1 ← DOOR MARK
1 ← HARDWARE GROUP

3. ALL DOORS 1¾" THICK UNLESS OTHERWISE NOTED
4. DOOR CONSTRUCTION:
   TYPICAL = SOLID CORE
   HC = HOLLOW CORE
   HM = HOLLOW METAL
   AL = ALUMINUM & GLASS
5. FACING & FINISH:
   TYPICAL = RED BIRCH, TRANSPARENT
   PT = PLASTIC LAMINATE, TEXTURED
   MP = METAL, PAINTED
6. GLASS:
   TYPICAL = CLEAR PLATE
   SG = SHEET GLASS
   CW = CLEAR WIRE
   TP = TEMPERED PLATE
   LG = LEAD GLASS
7. ¾, 1, 1½ ETC. INDICATES HOURS OF FIRE RATING.
8. TYPICAL FRAMES SHOWN "✓". NUMBER INDICATES DETAIL SHOWN ON SHEET_____.

SIMPLIFIED WINDOW SCHEDULE

The window and louver schedule is a summarized depiction of various units throughout the project. It should clearly identify types, dimensional aspects, and operational characteristics. For small and simple projects, where elevations can easily show all window types, no schedule is required.

RECOMMENDATIONS

1. Scale ¼ inch = 1 foot 0 inches.
2. Window wall construction should be indicated and detailed separately.
3. Unit Elevation should always be drawn as viewed from the exterior side.
4. Dimensions shown on schedule should be rough or masonry openings.
5. Directions of vent swings or sliding panels should be indicated on the schedule.
   a. Top hinged always swings out toward exterior, u.o.n.
   b. Bottom hinged always swings in toward interior, u.o.n.
   c. Side hinged always swings out toward exterior, u.o.n.
   d. Pivot hinged always rotates about indicated axis.
6. Indicate symbols on Floor Plans typically. For small projects keying to elevations may be preferable.
7. If a window type is placed into more than one type of wall construction, detail reference marks should be placed on the building elevations where each type of wall construction occurs.
8. If a window or louver unit is to be made of more than one material, a type number or letter should be assigned for each material type.
9. Typical glazing should be noted in the specifications. Special glass types and thicknesses other than typical or minimum, as required by referenced glazing standards, should be noted on the schedule.

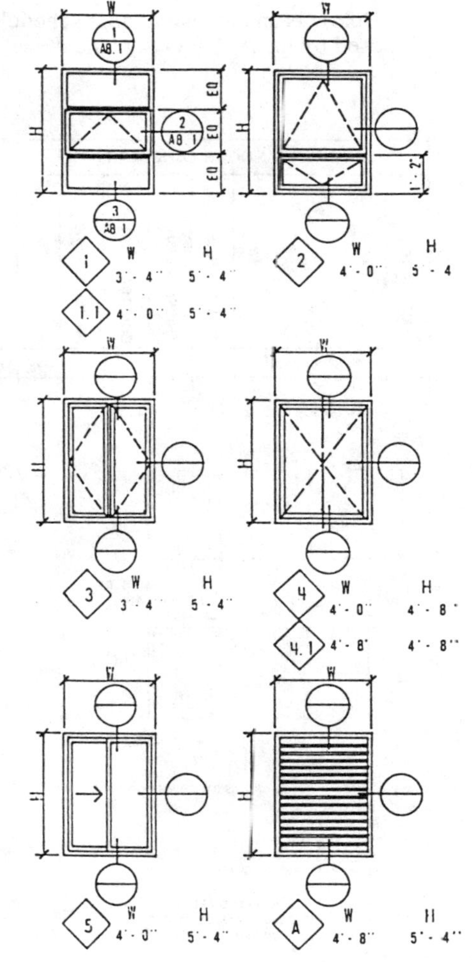

SCHEDULES

CONTENT CHECKLIST

BASELINE CONTENT

Baseline content is that which is required for building department approval and construction either by an owner-builder or through a negotiated contract with an owner-selected contractor.

__ Notes on plan identifying finishes, door types and sizes, and window types and sizes

STANDARD CONTENT

Standard content is that typically required for building department approval, construction cost estimating and bidding, and complete construction by a qualified general contractor.

Includes previously listed content plus:

__ Schedules as lists identifying generic substructure and finishes with references to specifications

__ Door and window symbols referenced to door and window frame schedules which are referenced in turn to detail keys showing sill, jamb, and head conditions

EXTENDED CONTENT

Extended content is that required for projects requiring extra design office attention because of complex construction, elaborate detailing, and/or extended collaboration with specialized consultants.

Includes previously listed content plus:

__ General note explaining schedule system

__ Photo details of textures

__ Photocopy illustrations of manufactured doors and windows

__ Hardware schedule with door schedule

__ Short-hand door and window schedules included on floor plan sheets for convenient contractor reference

Large, full sheet schedules are not necessarily more comprehensive, especially if they have highly repetitive elements. Simpler schedule formats that fully convey all materials combinations in each room, keyed to a simple symbol, are preferred.

Notes:

___

___

___

___

___

___

___

___

___

___

___

DOOR SCHEDULES

__ DOOR TYPES DRAWN IN ELEVATION
(1/2" scale is typical.)

__ SYMBOL AND IDENTIFYING NUMBER AT EACH DOOR

__ DOOR SIZES: __ WIDTHS __ HEIGHTS
__ THICKNESSES

__ DOOR TYPES: ___ MATERIALS __ FINISHES
(The number of doors of each type and size is sometimes noted.)

__ OPERATING TYPE: __ SLIDING __ SINGLE-ACTING
__ DOUBLE ACTING __ DUTCH __ PIVOT
__ 2-HINGE __ 3-HINGE

__ KICKPLATES

__ FIRE RATING IF REQUIRED

__ LOUVERS __ UNDERCUTS FOR VENTILATION

__ SCREENS

__ MEETING STILES

__ DOOR GLAZING __ TRANSOMS
__ BORROWED LIGHTS

__ DETAIL KEY REFERENCE SYMBOLS:
__ SILLS
__ JAMBS
__ HEADERS

__ MANUFACTURERS __ CATALOG NUMBERS
(If not covered in Specifications.)

__ METAL FRAMES:
__ ELEVATIONS
__ SCHEDULE
__ DETAILS

WINDOW SCHEDULES

__ WINDOW TYPES DRAWN IN ELEVATION
(1/2" scale is typical.)

__ WINDOW SYMBOL AND IDENTIFYING NUMBER AT EACH DRAWING

__ WINDOW SIZE __ WIDTH __ HEIGHT

__ WINDOW TYPE
__ DIRECTION OF MOVEMENT OF OPERABLE SASH AS SEEN FROM EXTERIOR
(Number of windows of each type and size is sometimes noted.)

__ GLASS THICKNESS AND TYPE

__ NOTE: __ FIXED __ OBSCURE __ WIRE
__TEMPERED __ DOUBLE GLAZING __ TINTED

__ SCREENS

__ DETAIL KEY REFERENCE SYMBOLS:
__ SILLS
__ JAMBS
__ HEADERS

__ MANUFACTURERS __ CATALOG NUMBERS
(If not covered in Specifications.)

__ METAL FRAME STOREFRONT OR CURTAIN WALL SYSTEM:

__ ELEVATIONS __ SCHEDULE __ DETAILS

FINISH SCHEDULES

__ ROOM NAME AND/OR IDENTIFYING NUMBER

__ FLOOR: THICKNESS __ MATERIAL __ FINISH

__ BASE: HEIGHT __ MATERIAL __ FINISH

__ WALLS: MATERIALS __ FINISHES
(Walls may be identified by compass direction code symbol if finishes vary wall by wall. Note waterproofing and waterproof membrane wall construction.)

__ WAINSCOT: HEIGHT __ MATERIAL __ FINISH

__ CEILING: MATERIAL __ FINISH

__ SOFFITS: MATERIAL __ FINISH

__ CABINETS: MATERIAL SPECIES AND GRADE
__ FINISH

__ SHELVING: __ MATERIAL __ FINISH

__ DOORS: MATERIAL __ FINISH
(If not covered in Door Schedule.)

__ TRIM AND MILLWORK: __ MATERIAL
SPECIES/GRADE __ FINISH

__ MISCELLANEOUS REMARKS OR NOTES

__ COLORS: STAIN AND PAINT

__ MANUFACTURER AND TRADE NAMES OR NUMBERS (If not covered in Specifications. Sometimes left for later decision with provision for paint allowance by bidders.)

__ EXTERIOR FINISHES:

__ EXTERIOR WALLS
__ SILLS __ TRIM
__ POSTS __ GUTTERS AND LEADERS
__ FLASHING AND VENTS __ FASCIAS
__ RAILINGS __ DECKING __ SOFFITS
(Included in Finish Schedule if not covered in Specifications.)

STANDARD DETAIL LIBRARY MANAGEMENT

The first rule of establishing a standard detail library is: Don't create details from scratch just for the library.

Plan on doing all future project construction details in such a form that the most generic of them can be filed for convenient retrieval, editing, and reuse. The only added investiment is taking the time to follow the uniform detail format and create a file number for finding and retrieving the detail later.

To make the job a little easier, use the Standard Detail Library file number index that follows.

THE ROLE OF CONSTRUCTION DETAILS

In terms of successful construction, details are the most crucial part of working drawings.

This isn't always understood or appreciated, so let's explain.

Working drawings convey information in sequence from the general to the particular, from the broadscope overall (small scale) to the narrowscope (large scale). They also proceed chronologically to some degree, from sitework to framing to details and finishes.

The spatial and formal information -- shapes, locations, sizes -- are best presented by drawings and photos. (Imagine trying to verbally convey the information of a floor plan.)

The specifications convey legal responsibilities, standards, products and materials, workmanship requirements -- information which cannot be easily presented graphically. (Imagine conveying a legal contract as a set of drawings with minimal text.)

So drawings and specifications are clearly mutually supportive -- each part of the documents does what the others cannot. Similarly mutually supportive are the broadscope small-scale drawings and the narrowscope large-scale drawings.

The smaller-scale drawings -- plans, elevations, cross sections -- are identifiers, graphic indexes, pointers, keyed maps that show generic materials and point the user to locations of more specific, larger-scale descriptions of construction.

Schedules may be a waypoint in the path of information, for example, when door and window types are shown on the plans and refer to door and window schedules. Then detailed construction of door and window frames are keyed to the schedules.

Similarly, a small-scale cross section may key wall sections, and wall sections will have keys to the large-scale details.

From the foregoing, it becomes clear that construction detail drawings contain the most crucial information and deserve the highest-level attention. This means they need to be managed.

While there's ample information available on materials, construction components, and essentials of detailing, there's little enough on managing details.

WHY STANDARD DETAILS?

There are two main arguments **for** using standard or reference details:

1) Construction details make up 50% or more of the drawings in many sets of construction documents.

2) A large percentage of construction details -- sometimes half or more -- are repeated in various projects.

Add up those two arguments and it seems you might cut out 25% of your drafting by systematically reusing repeat details instead of redrawing them each time. It's true -- potentially. But it doesn't usually work out. Here's why:

Standard details are often started without a complete long-range plan as to what should be standardized and why. One result is a file full of details of stock items traced from manufacturer's catalogs and other items that don't need to be detailed and will just clutter up the documents.

Standard details are sometimes initiated with impractical or expensive ways of transferring the details to new sets of drawings. Some offices have started detail books only to find that their local building code agency (unjustifiably) won't accept them.

Standard details in CADD files are sometimes abandoned, because it takes so long to look up potentially reusable details, many drafters just conclude it's faster to draw them from scratch.

Even if a system gets off the ground, someone has to watch over it, *manage* it. Most systems fall into obsolescence and disarray within their first year. Filing, storage, and retrieval systems disintegrate from misuse and neglect.

Then there is the problem of staff and/or management resistance: "This isn't supposed to be a factory," or "We do creative design -- no chance for standardization in our work." If that resistance comes from a high position in the office or otherwise has power behind it, it will be virtually impossible to create and maintain a successful system.

So, here is all that potential and here are all these barriers. How can you break through it all with a successful, long-lived system? The main steps are as follows:

1) Review old working drawings and identify repetitive details. Work out some numbers on what the actual savings will be for your firm if you establish a bank of reusable details. This step is essential for further planning and budgeting, and it's an important tool for dealing with skepticism or resistance.

(Creative people sometimes misunderstand how much design work is really repetitive. Most construction of any kind is highly standardized. No office invents a whole new structural system, flashing system, and roofing system, or unique wall construction, partitions, floors, fixture mountings, stairs, hatchways, and so on, for each new project. But it's often unclear how extensive the repetition really is. There's no substitute for counting details to find out.)

2) When reviewing repetitions of details from project to project, look too for *partial* repetitions. For example, many wall sections are alike in most ways and different in others. Curbs, flashing, slabs, etc., may have 90% similarity from job to job without being literal repeats. Plan on reusing the *common partitions* or *backgrounds* of such details that will likely repeat from project to project.

3) Prepare a list of all detail types that you would consider storing in your standard detail file. You'll use the list to plan and schedule the creation of your master detail file. When you have a list, you have a basis for gathering reference materials and sample details. You can keep the compiled materials in an "in-process" file, ready for people to work on whenever they're between assignments or otherwise on down time.

4) Plan to create all future details for each new job as if they were for the standard detail file *first.* This is a simple way to maintain the integrity of the detail system and ensure that it doesn't get bypassed and ultimately abandoned. You'll prepare *all* job details as if they were standards. Some will be unique and unusable on future work. They'll go in a special "reference section." Some will be of a type that might be reused but only in part. Those will be completed up to the point where they might have further use. They're frozen at that point as standards, copied for the file, and then continued with unique elements needed for the project at hand.

5) When preparing to create all future details as standards for the detail file, keep the variables as nonspecific as possible. For example, if creating a wall fixture attachment that would use the same screws, bolts, and so forth, regardless of whether the wall is masonry or concrete, don't show the wall construction or note it on the master. If an opening or material might vary in size in different projects, leave it undimensioned. The opening or material can be dimensioned as needed to fit specific jobs, and noted "N.T.S." (not to scale).

6) Classify and label some details as "in process," some as "reference," and some as "standard." That will help keep the files properly segregated. If there's considerable resistance and resentment toward standards. then call them all "reference details," and call the detail bank a "reference detail library." No one seriously objects to using detail references. It's done -- it has to be done -- all the time anyway.

7) Create a drafting guide of graphic standards for detail drawing. List standards that are necessary for all reprographics. This includes minimum lettering height: 1/8" high with 1/8" space between lines of lettering. Minimum minor title letters: 3/16" high, etc.

NOTE: Keep in mind that details, while prepared at different times on different sheets and put in different parts of the file, will ultimately be assembled together on the same large working drawing sheets. Consistency in composition, lettering, and so on, will pay off in coherence and readability of assembled detail sheets.

8) While doing preliminary planning and preparation for a detail bank, consider the following possibilities (but don't make immediate decisions on them):

-- If doing manual drafting, you may choose to draw standard and reference details freehand. This can be very economical and speedy. If final details or drawings sheets are photoreproduced, freehand wok is "straightened out" in the process.

-- You can use photographs of actual construction in standard and reference detail drawing. This is another easy, small-scale way to become accustomed to an important technique.

-- Coordinate your standards format with a modular, overall drawing sheet design. Both the popular size sheets -- 24" x 36" and 30" x 42" -- can be designed as composites of 8-1/2" x 11" units. Although standard details are prepared and stored on 8-1/2" x 11" sheets, that doesn't mean each detail will take up that much space in the final drawings. Your actual "detail module" will be some subunit of the 8-1/2" x 11" detail master sheets. Some firms use a 4-1/4" x 5-1/2" unit or multiples of it.

9) Plan on making each detail complete in itself, without cut lines. Avoid having parts of a detail on different 8-1/2" x 11" masters, unless they can be used as single details in themselves.

10) Select a filing indexing system. This deserves some careful thought -- it's another one of those decisions you'll be living with for a long time to come. Some firms use an adaptation of the CSI division system, so the detail bank complements the organization of specifications files. Some firms, including very large ones, use a very simple alphabetical system: doors are under "D," roofing under "R," for example. If your sets of drawings will be organized according to divisions that match the construction sequence, that might be a logical way to subdivide your standard detail file.

11) Be sure someone will be permanent overseer of the detail system. New details have to be processed, approved, and filed. Old details will have to be backchecked when there are complaints or suggestions from construction sites. Files will have to be periodically sorted and updated as items get misfiled or lost. New staff members will have to be shown how to use the detail system and how to prepare original details for it. Old staff will have to be reminded of all that occasionally. It's an ongoing process, and, except in some small offices, it won't take care of itself. A production assistant or production coordinator, or both, may be in charge. Or, if a traditional structure is maintained, maybe it should be watched over by the chief draftsperson, office manager, or specifications writer. Ideally, the person who conceives and implements the file should have ultimate responsibility for its upkeep.

12) Acquire basic storage equipment: a couple of file drawers, a set of hanging file folders, file-out cards to indicate materials that have been removed, and a three-ring "reference

binder" that will contain the index and photocopies of all details in the detail bank. (Some firms provide a complete master index three-ring binder to all drafting staff.)

13) Design the detail master sheet format to include the following reference information:

a. File or index number at upper right-hand corner, for easy identification in file folders.

b. Detail name, also at upper right, for easy visibility.

c. Detail scale, person who drew it, original drawing day, and revision dates.

d. A column to record where and when detail has been used and when last updated according to use. (If ever a detail presents a job-site problem, it can be backchecked through other jobs under construction or nearing construction.)

14) Design a preparation, filing, and retrieval system that will keep the file organized and under control. A recommended system is described further on.

Notes:

____

____

____

____

CONSTRUCTION DETAILS ARE THE KEY TO QUALITY CONTROL

Details are the focal point in the quality control programs of hundreds of U.S. design firms. There are three main reasons why:

1) "Most construction failures occur in the details -- either in design, drawing, or construction," says architect-engineer and building failures expert Ray DiPasquale. He makes the point that while engineering students, for example, learn moment diagrams and how to use shear and load tables, the main danger points are in the joints, the detailed connections. A knowledgeable Cambridge, Massachusetts, engineer quoted in *Engineering News Record* concurs: "A large percentage of the time, collapses seem to be related to details rather than large structural principles." He adds: "It's the odd detail that you don't do everyday that gets you into trouble." This doesn't mean you stop inventing new details. It just means taking care -- extra care -- through a systematic master detail system and a coordinated system of working-drawing project management and quality control.

2) Details, when coded by the CSI numbering system, can become an integral link that helps tie together every part of design and construction, including broadscope drawings, notation, schedules, specifications, and project decision-making documentation.

3) Besides quality control, architects and engineers achieve great time and cost savings from master detail systems. An innovative residential architect, for example, gets 30 to 50 percent of detail reuse from his files. He originally didn't expect any reusability because of the diversity of his practice and

the novelty of his designs. He learned, as have many others, that fundamental construction remains constant despite highly divergent design details. Thus, floor drains, roof drains, insulated drywall partitions, roof scuttles, metal ladders, cavity walls, etc., are found in only a very limited variety and are mainly repetitive from building to building. Larger-office users of standard or master detail files often report that 80 percent or more of the details on a complex detail sheet may come directly from the master files. That's an immense savings over time. And, it's a savings achieved by improving, not diminishing, design and construction quality.

THE ESSENTIALS OF A DETAIL SYSTEM

We've spent years researching detail systems and creating the beginnings of what has become a national master detail system, and there's one thing that emerges loud and clear: Most working drawing details are repeats. They're copies, to one degree or another, of previous details. Those that aren't copied are researched and redrawn from scratch, time and again in office after office. Whether your drafters are copying old details or reinventing them, it's a huge waste of time and money.

We'll elaborate on all the specifics of creating and running a master detail system shortly. First, here's a synopsis of the main components of such a system and how they come together.

First is a reference detail library collected from past jobs, other architectural and engineering firms, and published manuals and catalogs. Details are sorted and filed by category in a custom, in-house version of *Graphic Standards*. The details cover all kinds of construction, especially unusual, one-of-a-kind conditions. They're for help when the "one-of-a-kind" circumstance comes up again. Importantly, when you go to an integrated master detail system, you get multiple use of every single detail that will ever be drawn in the future. Those that aren't suited to a master or standard system for direct reuse are still at least useful in the reference file.

Second, in addition to reference details, is a file of generic or prototype details. These may be like our Guidelines detail starter sheets that include only major elements that are most repetitive and leave off the minor items most likely to vary from job to job. Ed Powers of Gresham & Smith sends their new generic details to construction-related trade associations for review and comment. That's an outstanding effort to record the very best construction information available and make it an integral part of design documentation.

The third quality control tool: master details. These are more specific details, mostly originated for ongoing projects by elaborating on starter sheets or generic details. They're left partly undone, copied, and the copies later reviewed for inclusion in the master detail file. If they don't quite cut it for the master file, they go to the reference file.

The fourth tool, a jobsite feedback form, keeps the detail system fully up to date. If there's a troublesome detail discovered on a project, project representatives are required to send a brief memo back to whoever is in charge of the quality control-master detail system. The memo names the detail-type file number (which is printed with title and scale with every detail used in the drawings). The system manager looks up the detail, and, while identifying the problem, looks at the detail drawing's detail history form. That's a space on the detail format sheet that identifies where else and when a detail has been used. The combination of jobsite feedback form and detail history form uncovers problems, allows quick corrective action on other projects that may be using the detail, and constantly helps in screening and upgrading the entire master file.

There's more on all these points in the text that follows. The Appendices of this book include comprehensive detail file number indexes for filing and retrieving construction details. The numbers are coordinated with the CSI Masterformat, and you'll find them an extremely useful timesaver as you establish or update your own master detail file.

THE DIFFERENCE BETWEEN REFERENCE DETAILS AND STANDARD DETAILS

Reference details are those you might review for guidance in designing for some unfamiliar construction situation. They might be from previous jobs, from other offices, from product manufacturers, or from guidebooks such as *Graphic Standards* or *Time Saver Standards.*

Most offices don't keep a formal reference detail file. They may keep useful details from previous work, but the details are usually not filed or indexed in any systematic way. Systems drafting-oriented offices use reference details increasingly these days. They're an excellent supplement for any master or standard detail file.

Standard details represent prevailing conventional construction. They show the repetitive conditions and are designed to be used directly -- with or without revisions -- in working drawings. Standard details commonly include sitework conditions, door and window details, common types of wall construction, and connections of manufactured items such as roof drains and metal ladders. Even small offices sometimes compile and use as many as 3,000 details showing common construction situations.

Besides being a major time and cost saver, standard details allow you to create and store the very best of construction practice. Offices that maintain well-managed detail files report a noticeable decline in problems at the jobsites. And there are definitely fewer problems with buildings after construction.

Besides being a major time and cost saver, standard details allow you to create and store the very best of construction practice. Offices that maintain well-managed detail files report a noticeable decline in problems at the jobsites. And there are definitely fewer problems with buildings after construction.

Keep in mind the difference between reference details and standard details. Reference details are a source of data in creating new original data. Standard details are directly reusable drawings of common repeat items of construction.

CREATING THE DETAILS

People don't have to be taken off other work to create an office master detail file. You can create a master file by doing all future details in a format that allows them to be easily used and entered in the master file.

Details created in this fashion are drawn as usual but in special format on 8-1/2 by 11 inch sheets. All details are later reviewed for potential reusability in future projects. Even if a detail isn't acceptable as a standard detail, it can still most likely be useful as part of the reference detail file. Either way, you gain additional usage and value from most details long after they've been drawn for their particular project.

If you follow the recommended standard format, you'll be creating your future details within an approximate 5-3/4 inch high by 6-inch wide "window" on 8-1/2 x 11 inch format sheets. The format sheets; will show how to size and locate the detail title, how to align lettering, etc. When everyone uses the same format, details assembled on carrying sheets will look consistent.

The general rule in creating new details for future projects will be to split the drawing process into two steps:

1) Bring the detail to a point of *near* completion. "Near completion" means to leave off the material indications, dimensions, and notes that might vary in different circumstances. Some opening sizes or fabrication sizes might be variable, for example, so they're not specified at this step. The idea is to avoid putting so much information on the detail that it loses its potential reusability.

2) After creating the detail up to a point of potential reuse, make two copies. Proceed to finish up your work for the job at hand on one copy. Add the variable material indications, dimensions, and notes that are necessary to complete the detail. The other copy is to review later for its potential use in the reference or master detail files. File the original separately. If you select a detail for the master file, the separate original provides backup- insurance against possible loss of the file copy.

Each detail accepted for either the reference file or for the standard detail file will receive a file number. This number will be integrated with the office's master specification numbering system. File numbers, and a master file number index, provide slots or "address numbers" for all filed details. It makes for the most convenient retrieval and cross-referencing of details after they've been filed away.

FINDING AND RETRIEVING DETAILS FROM THE STANDARD FILE

As just described, each filed detail receives a permanent address or file number. The number identifies the detail's construction division relative to specifications. The detail may be numbered according to dominant material or by function. The specification number system determines exactly what numbers to use to identify the major divisions and their subcategories, or the broadscope and narrowscope. (See the Detail File Index later in this text.)

The division numbering is reflected in CADD files or a physical detail file.

A "physical" file consists of file drawers containing hanging "Pendaflex"-type folders. Each folder is marked with a specification division name and broadscope number.

"Physical" paper files or CADD files follow the same system. Within each folder are smaller folders that contain the subcategories of details within a particular broad division. For example there might be a large folder identified as "Site Work-Division 2." Within that might be individual folders for "Curbs," "Parking Bumpers," etc. Each of those folders has a CSI-related number. Then the details within a folder each have their single identifying number. Thus a detail might be numbered 02528-4. Following the CSI format, that number would mean "concrete curbs within Division 2 site work." And the number following the hyphen would mean that this detail is the fourth concrete curb detail in the file.

It would be inconvenient to poke through a lengthy index of detail names and numbers to find a detail that you hope might be on file, so you need a convenient cross-reference sys tem -- one that shows what details are available and tells how to find them in the file.

The cross-referencing, or "lookup," system is in the form of a computer index or three-ring binder -- a detail catalog. The detail catalog is a filing system ancillary to the file number structure of the file folders. Instead of being shown in the catalog in the way they're filed, details are shown in the sequence that people would be most likely to look for them. If you want to check out some window details, instead of looking through separate divisions of steel windows, aluminum windows, and wood windows -- all separate specification sections -- you'd look in the general category of the catalog labeled "Exterior Walls." Then you'd search for the alphabetical subsection under "W" for "Windows." Within that you could look for further subdivisions of window types and materials.

When you locate some details in the catalog to try out, note their file numbers and have the masters retrieved and copied for you. Add special data unique to your project to complete the details, and proceed with check prints.

Notes:

____

____

____

____

____

TYPICAL ARGUMENTS AGAINST STANDARD DETAILS

"None of our buildings are alike; it isn't worth the time to reuse a few details from time to time."

Those who use standard detail libraries report a consistent 60% to 80% reuse factor on all their projects. That is, no matter how diverse their building types and designs, as much as 80% of all details used often come directly from the detail library. That's because no one invents whole new roof drains, sliding doors, grab bars, suspended ceilings, curbs, etc.; they come out of boxes and are combined with equally standard roof, floor, and wall construction.

Imagine what 60% to 80% reuse means financially. A typical detail will take two to four hours to design, sketch, draft, check, correct, and recheck. If your average drafting time costs $12 direct hourly pay, plus 30% benefits, for a total of $15.60 per hour, then a medium-size project with 40 detail drawings at 3 hours per detail costs $1,872. If you get just 70% reuse of details on the same project, the savings is $1,310. Multiply that by 12 projects a year for an annual potential savings of $15,724. After a few years of that, you're talking substantial money.

"We don't have the time to sit down and draw all those details."

You'll be drawing all those details on future jobs anyway, so why not start by making the next detail you draw for the standard file as well as the job? Then the next . . . and the next. Eventually you'll have a complete file.

"But most details are unique; they won't be reused again."

TYPICAL ARGUMENTS AGAINST STANDARD DETAILS continued

Maybe. Actually, most users are pleasantly surprised to find a much higher reuse factor than they expect. Besides, those details that don't end up in a totally reusable Standard Detail File can go into an auxiliary Reference Detail File. You may only do a residential swimming pool detail once in a blue moon, but you can be sure there will be a need for related information some day, so why not keep it for possible future reference?

"It's too much trouble to draw them on small sheets, file them, find them. . . ."

Details drawn on small sheets (8-1/2" x 11") are faster to draft than those drawn across back-stretching, full-size drawings. And with the commonly-used enlargement reduction copiers and large-format copiers, it's never been easier, faster, or cheaper to handle copying, pasteup, and reproduction. Not only are half of most details on a job reused on a project, the process of pasting up new and old details is markedly faster than drafting them all in the traditional manner.

"We have CADD, and standard details are still hard to manage."

Computers are good for storing standard details, but it's time-consuming and wasteful to plot them out on full-size sheets. If you're using a pen plotter, it's faster to print details on small sheets with high resolution dot matrix or laser printers, and manually cut out and paste them up. That makes the inevitable changes easier to manage, too.

"You can't possibly make enough details for all the different combinations of construction -- like wood windows in wood frame walls, in brick walls, block walls. . . ."

A few hundred standard detail components can give you a library of tens of thousands of possible detail combinations. For example, just thirty drawings -- including ten window sections, ten sill types, and ten differnet wall types -- will give you the raw material for one thousand possible detail drawings. You can quickly recombine them by tracing, by combining transparency overlays, by paste-up, or on CADD.

STANDARD DETAIL SHEETS

Certain kinds of details may be treated as standard sets or clusters rather than individual unrelated drawings. Sitework details, for example, often repeat themselves as a group from project to project.

Another, more obvious example is the "nomenclature and symbols" sheet that shows the abbreviations, standard notes, symbols, etc., that are office standards for all projects. Such sheets can be prepared just as standard details would be, except they're filed as full-size drawings, rather than units of 8-1/2" x 11".

Other drawing types that may be prepared as full-size sheet standards include: cabinet work, wall construction, door frame schedules and details, and integrated partition and ceiling systems.

Every discipline has its group of consistently reusable elements. Many structural engineers, for example, prepare an "S-1" drawing that contains all the most commonly repeated structural details.

A very important type of "standard" sheet is the "interior elevation fixture standards." If your firm does any particular building type repeatedly, there'll be a set of fixtures, housing cabinetry, plumbing fixtures, library fixtures, penal institution fixtures, etc.

Often drawn are interior elevations that really only have the function of showing the heights and, perhaps, appearance of wall-mounted fixtures. Since the *horizontal* locations of fixtures such as drinking fountains, phone booths, fire hose cabinets, and so on, are shown on plan, all that's needed is the height indication and possibly some detail key reference. That can be provided in a standard elevation drawing showing the fixtures, their heights, sizes, appearance, and any necessary fixture mounting detail keys.

Notes:

____

____

____

____

____

____

____

____

STANDARD DETAIL BOOKS

Standard detail systems are most often described in terms of assembling details on large working drawing sheets. The following question inevitably arises: If you're keeping your standard masters on 8-1/2" x 11" format sheets, why not just use them at that size directly in detail books?

That's definitely more convenient, and many offices do their details in book form. Sometimes they're incorporated with the bound specifications; sometimes complete working drawings, details, and specs are all bound together as one book-sized unit.

There is controversy about standards books. Many contractors say they're inconvenient to use and easy to lose (or ignore) on the job. They also complain that they don't know if details included in such books are *real* and actually apply to the job as shown. They've seen book details that were just randomly-applied boilerplate, with only the vaguest application to the job at hand.

On the other hand, some contractors and subcontractors say they're extremely handy. One fabricator told us: "It saves me and my workers a lot of time to be able to pull a page and carry it right up the ladder instead of wandering back and forth to the print table."

In general, if the drawings are applicable to the job, and if the addressing and referencing system is very simple and clear, there needn't be much objection to using detail books.

Detail books can be improved by including reduced-size prints of key plans, elevations, and sections. That helps people to see the isolated details in context.

CHOOSING SYSTEMS AND FORMATS

You've seen several types of filing/retrieval systems and format styles in this text -- all used successfully by architects across the U.S.

Take care in choosing or revising your own system; you'll be living with it for years to come.

Many otherwise useful standard detail libraries have been abandoned and became obsolete through disuse, because they were too complex or too simple for the evolving office practice.

Review all indexes, formats, and other text with key people in the office -- production supervisor, project managers, senior drafters -- and get their input. Then use your best judgment, and decide.

Then, set up strict rules that all future details, no matter how unique, be done in a consistent format, and, when filed, be given appropriate file numbers by you or whoever is in charge of the system.

Which brings us to the final lesson learned in office after office: Someone has to be in charge of the system. The system won't run itself, drafters will not follow the format rules unless reminded to do so periodically, and designers and project managers will tend to make up new details from scratch, rather than reuse the same details from the detail library.

To gain all the significant economies, efficiencies, and quality control benefits of the system, it has to be easy to use, agreeable to the users, and above all, it must be enforced.

STANDARD CONSTRUCTION DETAILS PROCEDURAL CHECKLIST

Procedures for creating details and managing a standard detail library still differ considerably from office to office, but here are the basics, widely tested and accepted procedures and standards for creating construction details.

RULES FOR DETAIL DESIGN

__ First list details and detail types, as part of a miniature mock-up working drawing set. Draw cartoon miniatures of plans and elevations, then indicate bubble keys at joints, junctures, and typical sections where details will be needed.

__ Make miniature cartoons of needed details on one-fourth size mock-up sheets, to show their approximate size, positioning and coordination with one another. Name the details.

__ Search your standard or master detail file for possibly usable details for the project at hand. Make office copier prints and assemble them on a paper carrier sheet as a rough checkprint.

__ Search your reference detail library for details that will aid in the design of exceptional and non-standard details. Assemble office copier prints onto carriers for review.

__ Search the office technical library for published details, such as product manufacturer and trade association literature and drawings.

__ Proceed with rough design sketches on 8-1/2" x 11" sheets as necessary to develop new original details. Copy the wholly new details and assemble the checkprint copies on carrier sheets.

__ Use red pencil or pen to show revisions, additions, and deletions on the full-size detail checkprint sheets.

This essentially completes the research and planning groundwork.

SYSTEMATIC PRODUCTION

__ When creating new details, such as wall sections, watch for opportunities to use common background data as base sheets and overlays or CADD layers, to show the unique detail situations in context.

__ Watch for opportunities to use oversized, freehand-sketched details as photo reductions, instead of redrafting them from scratch.

__ Use a standardized 8-1/2" x 11" detail format sheet identical to your standard detail file format sheets for ALL new detail final drawing. Remember, new details that aren't suited for direct reuse in your standard file are still reusable as sources of ideas and technical information in a separate reference detail file.

__ Set details within a consistent cut-out/paste-up "window" (6" wide by 5-3/4" high).

CONSTRUCTION DETAILS PROCEDURAL CHECKLIST
continued

DRAFTING STANDARDS

Besides seeking general sharpness and high contrast, contemporary design and working drawings have to be designed for special reproduction processes. Elements may be photo-reduced, for example, so elements in the original drawing must be extra large, to remain readable. Or drawing components may be copied through several generations of reproduction, so they have to be extra crisp, clear and black to begin with, or they'll fade away.

__ All lettering should be a minimum of 1/8" high, with 3/32" to 1/8" spacing between the lines of lettering. Major titles should be a minimum 1/4" high letters, minor titles minimum 3/16" high.

__ Line work must be consistently black and vary only by width, not by "darkness" or "lightness." "Light" or grey lines disappear during printing.

__ All symbols such as arrowheads, feet/inch marks, circles, etc. must be large and unmistakably clear. Small symbols, small circles, etc. tend to "clog up" during reproduction.

__ The lines in work or cross hatching must be spaced at least 1/16" apart. If lines and patterns are too close together, they'll run together in the reproduction process.

__ It's not likely any more, but just in case: No hand poche' should be applied on the back of a drawing sheet.

__ Don't let line work touch letters or numerals. Don't let fractional numbers touch their division line. Otherwise they tend to blur together.

CONSTRUCTION DETAILS PROCEDURAL CHECKLIST
continued

GRAPHIC PROCESS

__ Draw from the general to the particular and from the substructure or structure outward to the finish materials.

__ Draw in layers, so that a detail drawing is substantially a completed composition at any point along the way.

__ Draw in sequence from the primary to the secondary: detail component first, size dimensions second, notation third. Add textures, cross hatching and profiling last.

__ In general, keep the exterior face of construction facing to the left, interior to the right.

NOTATION

__ The primary purpose of notation is to identify the pieces of the detail by generic material name and/or construction function. Leave all extensive material descriptions, brand names, tolerances, and construction standards to specifications. On a small job that combines specification information with notes, follow the above rule, but add general or "assembly" notes, as a short-form specification on the drawing sheets.

__ In general, the first part of a note states the size of the material or part, tells material or part name second, and names position or spacing third. If using keynoting, the keynote legend will identify the names of materials or parts, and the size and sometimes spacing will be tagged onto the keynote reference bubble in the field of the drawing.

__ In general, notation should be in a column down the right-hand portion of the detail window. Locate notes elsewhere if necessary to avoid crowding or enhance clarity.

__ Follow standard nomenclature and abbreviations. The office should follow professional societies' recommended standards and print such guides as part of General Information.

LEADER LINES

__ Preferred line style is straight-line, starting horizontally from the note and breaking at an angle to the designated material or detail part.

__ Still use curved leader lines. They tend to create a cluttered appearance.

__ Don't use leader lines from a note to more than one detail or section.

DIMENSIONING

__ Avoid fractions where possible. When using fractions, never use those that can't be measured with normal scales.

__ Don't draw dimension lines directly to the face of a material. Unless you have no other option, draw only to lines extended from the material or detail part.

__ Add terms such as "Hold," "Min.," "Max.," "N.T.S.," "Equal," "Varies," etc., to clearly state your intent on the accuracy or absoluteness of your dimensions.

__ Avoid double dimensions that show a size dimension on one side of the detail and repeat it on the other side.

CONSTRUCTION DETAILS PROCEDURAL CHECKLIST
continued

LINE WIDTHS

__ Break lines, dimension lines and cross hatching are narrowest. General background construction lines are medium width. The primary detail object should be "profiled" with the widest line. Equivalent pen tips would be: 000 or 00 for narrow lines, 0 to 1 for medium, 1,2 or 3 for profile lines. Exact choice of line widths is variable, depending on the scale of the detail. Wider lines look excessive in small-scale details, and the lesser line widths look weak with the larger scale details.

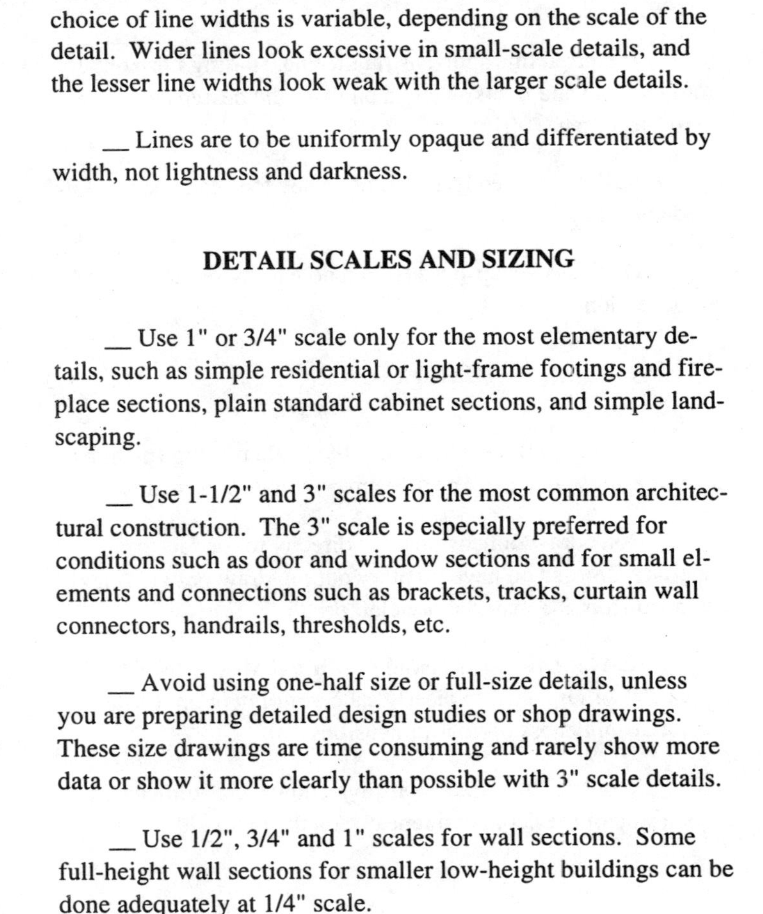

__ Lines are to be uniformly opaque and differentiated by width, not lightness and darkness.

DETAIL SCALES AND SIZING

__ Use 1" or 3/4" scale only for the most elementary details, such as simple residential or light-frame footings and fireplace sections, plain standard cabinet sections, and simple landscaping.

__ Use 1-1/2" and 3" scales for the most common architectural construction. The 3" scale is especially preferred for conditions such as door and window sections and for small elements and connections such as brackets, tracks, curtain wall connectors, handrails, thresholds, etc.

__ Avoid using one-half size or full-size details, unless you are preparing detailed design studies or shop drawings. These size drawings are time consuming and rarely show more data or show it more clearly than possible with 3" scale details.

__ Use 1/2", 3/4" and 1" scales for wall sections. Some full-height wall sections for smaller low-height buildings can be done adequately at 1/4" scale.

JOBSITE FEEDBACK FORM

Some offices require employees to carry cameras to jobsites with them to record any and all problems. Aside from project management issues, this has proven to be a great educational resource for design and technical staff members.

For the same purpose, some offices also use Jobsite Feedback Forms such as the one shown on the next page. This provides a way of recording construction problems but it's especially useful for upgrading the construction detail library.

JOBSITE FEEDBACK FORM

(This form requires an 8-1/2" x 11" sheet, your letterhead, and any necessary items from your standard submittal form.)

Jobsite feedback submittal to:

Copy to:

Date:

Job name:

Observer's name:

If problem involves specifications, note division number:

If problem involves large-scope working drawings, note sheet number:

If problem involves construction detail(s), note:

Job sheet number

Detail location number on the drawing

Detail file number (if any)

Description of design or construction problem:
(Use back of sheet if you need more space,
and add photograph if possible.)

Possible or likely cause of the problem:
(Use back of sheet if you need more space.)

STANDARD DETAIL LIBRARY FILING SYSTEMS

SIMPLIFIED ALPHABETICAL DETAIL FILING SYSTEM

Often a simple alphabetical system is perfectly adequate for the filing and retrieval of standard details.

A Master Manual
B Building code and zoning data blanks, drawing numbers
C Curtainwall, windows, entrances, store front
D Door and frame details and elevations
E Exterior details
F Finish schedule, door schedule
G Graphic symbols, abbreviation
H Handrails, stairs, ramps
I Interior details
M Millwork
P Partition-type sections
S Site-work details
T Toilet details and mounting heights

"A" & "B" are reference sections. Following is a list of one office's details filed under the alphabetical section titles from "C" on:

C Curtainwall, Windows, Entrances, Storefront Details

1. Typical aluminum entry doors
2. Sliding glass doors
3. Typical aluminum window sections
4. Brick sill sections
5. Sloping "greenhouse"-type skylight
6. Concealed overhead closer mullion
7. Drive-up bank-teller window
8. Wood guard railing on entry doors and sidelights

D Door and Frame Details, and

1. Typical hollow metal frame sections
2. Typical hollow metal and wood view panel
3. Door louver detail
4. Door and frame elevation sheets
5. Hollow metal borrowlight frame
6. Wood doorframe and sidelight frame
7. Overhead door and track
8. Roll-up grille and door
9. Pocket door detail
10. Door thresholds
11. Penthouse door sills
12. Elevator doors
13. Masonry block door lintel types

SIMPLIFIED ALPHABETICAL DETAIL FILING SYSTEM continued

E Exterior Details

1. Typical roof parapets
2. Typical gravel stops
3. Typical window head flashing
4. Typical masonry wall flashing blowup
5. Roof expansion joint curbs
6. Brick control joint detail
7. Bank or book depository
8. Roof scuttle
9. Prefab metal canopy
10. Wall louver details
11. Roof dome skylights
12. Loading dock details
13. Metal fence and gate
14. Roof drain and sleeve
15. Roof flashing
16. Roof equipment screen
17. Service yard screen
18. Foundation
19. Wall-mounted roof ladder

H Handrails, Stairs, Ramp Details

1. Wall-mounted brackets
2. Various typical railings
3. Tread-riser detail
4. Ships' ladders
5. Typical stair stringers
6. Areaway railing
7. Balcony railing
8. Ramp railings and slopes
9. Metal spiral stairs

I Interior Details

1. Base types
2. Divider strips and saddles
3. Folding partition details
4. Ceiling expansion joint detail
5. Brick column plan section
6. Fire extinguisher mounting height elevation, lettering, and section
7. Drapery track alcove
8. Strip electric heater mounting
9. Partition expansion joint
10. Floor expansion joint
11. Light cove detail
12. Wall-mounted light strip
13. Movable panel partition track and jambs
14. Toilet room soffit light
15. Door mat recess and frame
16. Typical lay-in ceiling details
17. Chalkboard and tackboard
18. Directory mounting detail
19. Hospital corridor rail
20. Drinking fountain heights
21. Drywall column furring plan
22. Acoustic ceiling coffer
23. Kitchen equipment base

SIMPLIFIED ALPHABETICAL DETAIL FILING SYSTEM continued

M Millwork Details

1. Cabinets and counters
2. Open shelving and adjustable standards
3. Coat rod and hat shelf section
4. Projector window and shutter
5. Interior borrowlight details (wood)
6. Coat hook strip mounting
7. Wood grille ceiling or soffit
8. Projector screen cabinet
9. Reception counter

S Site-Work Details

1. Walk expansion and control joint
2. Railroad tie retaining wall
3. Bollard details
4. Trench drain detail
5. Bike rack detail
6. Concrete curb detail
7. Concrete drive section
8. Concrete walk section
9. Brick paving
10. Asphalt curb detail
11. Asphalt paving section
12. Parking light standard base
13. Flagpole base
14. Bank drive-up island and culvert
15. Pipe guard for driving
16. Landscape planters
17. Traffic island
18. Concrete tire stop
19. Concrete or stone bench
20. Paving pattern plan

T Toilet Room Details and Heights

1. Toilet room accessory heights
2. Handicap rail and fixture dimensions
3. Toilet partition bracing above ceiling
4. Lavatory counter section
5. Mirror mounting trim
6. Shower receptor detail
7. Recessed towel and wastepaper wall section

STANDARD DETAIL INDEX BY CSI NUMBER

The checklist that follows this index will identify all the likely details you'll need to include in working drawings.

The list below serves a different purpose. It lists general detail types in sequence according to their CSI coordinated detail file numbers to aid you in filing and reusing details in a standard or reference detail library.

DIVISION 1 **(Not applicable in this list.)**

DIVISION 2 **SITEWORK**

02200 EARTHWORK

02350 PILES

02400 DRAINAGE
- 02410 Subdrainage and Foundation Drain
- 02420 Surface Run-off (Area Drain, Storm Drain)
- 02431 Catch Basin
- 02435 Splash Block

02440 SITE IMPROVEMENTS
- 02444 Fencing, Chain Link
- 02446 Fencing, Wood
- 02447 Planter
- 02451 Guard rail (Bollard, Stancheon)
- 02452 Signage
- 02455 Signage
- 02455 Parking
- 02457 Bicycle Rack
- 02460 Play Equipment
- 02470 Site Furniture
- 02477 Shelter (See also, Protective Covers 10530)

02480 LANDSCAPING

STANDARD DETAIL INDEX BY CSI NUMBER continued

02500 PAVING AND SURFACING
- 02515 Concrete Paving
- 02516 Asphalt Paving
- 02528 Concrete Curb
- 02577 Paving Marking

02600 PIPED UTILITY MATERIALS AND METHODS
(Includes manholes, cleanouts, hydrants, etc.)

02700 PIPED UTILITIES
- 02721 Storm Sewerage
- 02722 Sanitary Sewerage
- 02730 Water Well
- 02743 Septic Tank

02800 POWER AND COMMUNICATION
(Includes towers, poles, underground lines, etc.)

DIVISION 3 CONCRETE

03250 CONCRETE ACCESSORIES
(Includes anchors, joints, waterstops.)

03300 CAST-IN-PLACE CONCRETE
(Includes retaining walls, piers, footings, walls, columns, beams, etc.)

03400 PRECAST CONCRETE
(Includes deck, plank, beams, tilt-up, etc.)

03500 CEMENTITIOUS DECKS
(Includes gypsum deck and plank, concrete, wood fiber, asphalt and perlite.)

DIVISION 4 MASONRY

04150 MASONRY ACCESSORIES
(Includes anchors, control joints, etc.)

04200 UNIT MASONRY
- 04201 Cavity Wall (Brick and Concrete Block)
- 04210 Brick
- 04212 Adobe
- 04220 Concrete Unit Masonry (H.M.U.)

04400 STONE

STANDARD DETAIL INDEX BY CSI NUMBER continued

DIVISION 5	**METALS**
05100	STRUCTURAL METAL FRAMING
05300	METAL DECKING
05500	METAL FABRICATION 05510 Metal Stair 05514 Exterior Fire Escape 05515 Ladder 05521 Stair Handrail 05530 Grating 05531 Floor Plate
05700	ORNAMENTAL METAL (Includes ornamental stairs and handrail, prefab spiral stair, and ornamental sheet metal.)
05800	EXPANSION JOINTS
DIVISION 6	**WOOD AND PLASTIC**
06050	FASTENERS AND SUPPORTS
06100	ROUGH CARPENTRY
06130	HEAVY TIMBER CONSTRUCTION
06200	FINISH CARPENTRY (MILLWORK)
06400	ARCHITECTURAL WOODWORK 06410 Cabinetwork (Counters, Shelving, Wardrobe) 06420 Paneling 06430 Stairwork
DIVISION 7	**THERMAL AND MOISTURE PROTECTION**
07100	WATERPROOFING
07500	MEMBRANE ROOFING
07570	TRAFFIC TOPPING
07600	FLASHING AND SHEET METAL 07602 Wall Flashing
07620	SHEET METAL FLASHING

STANDARD DETAIL INDEX BY CSI NUMBER continued

07800	ROOF ACCESSORIES 07810 Skylight 07820 Stack 07822 Drain 07824 Curb 07830 Hatch 07870 Pitch Pocket
DIVISION 8	**DOORS AND WINDOWS**
08100	METAL DOORS AND FRAMES
08300	SPECIAL DOORS (Includes fire, coiling, folding, & overhead doors.)
08400	ENTRANCES AND STOREFRONTS
08500	METAL WINDOWS
08600	WOOD AND PLASTIC WINDOWS
08800	GLAZING
DIVISION 9	**FINISHES**
09100	METAL SUPPORT SYSTEMS 09110 Non-load Bearing Wall Framing System 09120 Suspended Ceiling
09200	LATH AND PLASTER 09202 Gypsum Lath and Plaster 09203 Metal Lath and Plaster
09250	GYPSUM WALLBOARD (DRYWALL) 09260 Gypsum Wallboard (Drywall) Wall 09270 Gypsum Wallboard (Drywall) Ceiling
09300	TILE
09400	TERRAZZO
09500	ACOUSTICAL TREATMENT 09510 Acoustical Ceiling 09520 Acoustical Wall
09550	WOOD FLOOR
09600	STONE AND BRICK FLOOR

STANDARD DETAIL INDEX BY CSI NUMBER continued

DIVISION 10	SPECIALTIES
10100	CHALKBOARDS AND TACK BOARDS
10150	COMPARTMENTS AND CUBICLES 10160 Toilet Partition 10162 Urinal Screen
10200	LOUVERS AND VENTS
10240	GRILLES AND SCREENS
10260	WALL GUARDS
10265	CORNER GUARDS
10270	ACCESS FLOORING
10290	PEST CONTROL
10300	FIREPLACES AND STOVES
10350	FLAGPOLES
10400	IDENTIFYING DEVICES (Includes bulletin boards, signs.)
10500	LOCKERS
10520	FIRE EXTINGUISHERS, CABINETS AND ACCESSORIES
10530	PROTECTIVE COVERS (Includes walkway cover, car shelter, awnings.)
10550	POSTAL SPECIALTIES (Includes mail chute, mail boxes.)
10600	PARTITIONS (Includes folding, demountable, mesh partitions.)
10670	STORAGE SHELVING
10700	SUN CONTROL DEVICES (EXTERIOR)
10750	TELEPHONE ENCLOSURES
10800	TOILET AND BATH ACCESSORIES

STANDARD DETAIL INDEX BY CSI NUMBER continued

DIVISION 11	**EQUIPMENT**
11010	MAINTENANCE EQUIPMENT
11020	SECURITY AND VAULT EQUIPMENT
11040	ECCLESIASTICAL EQUIPMENT
11050	LIBRARY EQUIPMENT
11060	THEATER AND STAGE EQUIPMENT
11100	MERCANTILE EQUIPMENT
11200	VENDING EQUIPMENT
11130	AUDIO-VISUAL EQUIPMENT
11150	PARKING EQUIPMENT
11160	LOADING DOCK EQUIPMENT
11170	WASTE HANDLING EQUIPMENT (Includes trash chute, incinerator.)
11400	FOOD SERVICE EQUIPMENT
11450	RESIDENTIAL EQUIPMENT 11451 Kitchen Equipment 11452 Laundry Equipment 11454 Disappearing Stair
11460	UNIT KITCHENS AND CABINETS
11470	DARKROOM EQUIPMENT
11480	ATHLETIC, RECREATIONAL, AND THERAPEUTIC EQUIPMENT
11800	TELECOMMUNICATION EQUIPMENT

STANDARD DETAIL INDEX BY CSI NUMBER continued

DIVISION 12	**FURNISHINGS**
12100	ARTWORK
12300	MANUFACTURED CABINETS AND CASEWORK
12500	WINDOW TREATMENT (Includes drape tracks & shades.)
12600	FURNITURE AND ACCESSORIES
12670	RUGS AND MATS
12700	MULTIPLE SEATING
12800	INTERIOR PLANTS AND PLANTING
DIVISION 13	**SPECIAL CONSTRUCTION**
13070	INTEGRATED CEILINGS
13080	SOUND, VIBRATION CONTROL
13120	PRE-ENGINEERED STRUCTURES
13140	VAULTS
13150	POOLS
13410	LIQUID AND GAS STORAGE TANKS
13980	SOLAR ENERGY SYSTEMS
13990	WIND ENERGY SYSTEMS
DIVISION 14	**CONVEYING SYSTEMS**
14100	DUMBWAITERS
14200	ELEVATORS
14300	HOISTS AND CRANES
14400	LIFTS
14500	MATERIAL HANDLING SYSTEMS
14700	MOVING STAIRS AND WALKS

ERRORS & OMISSIONS IN WORKING DRAWINGS

This may surprise you, but working drawings that were done around 1900 have essentially the same problems as those we do now (as do those done in the 40's, the 60's, and the 80's.

Old drawings often LOOK better. Ink and linen line quality was pleasing, ornament was abundant, and lettering more meticulous and less rushed. But, in general, the average drawings of yesteryear had about the same number and kinds of errors and omissions as any set of drawings you would find today.

For example:

-- Nearly half of all working drawing sets have dimensioning gaps or errors, such as:

-- Missing:

-- Rough-opening size dimensions or notes.

-- Vertical dimensions of sills and lintels.

-- Location dimensions for floor drains in slabs (plumbing and architectural drawings often conflict visually, and when neither has dimensions, the plumber has to guess).

-- Cumulative dimensions to structural columns or grid for locating walls and partitions (the building frame will be there when walls are built, so structural dimensions are often ignored in architectural dimensioning).

ERRORS & OMISSIONS IN WORKING DRAWINGS continued

-- Seventy percent of plans for larger buildings have parallel strings of plan dimensions that don't add up or don't match structural frame dimensions. This is a further result of failing to reference walls and partitions to structural grids or to use cumulative framing reference points.

-- Over 40% of plan and elevation drawings include unrealistic fractions of inch dimensions such as 1/4", 1/8", and 1/16". Sometimes there are highly detailed studies of storefront mullions so tightly dimensioned that only a machinist could make them.

-- Over 55% of drawings lack clear reference points for height dimensions in exterior elevations. The datum start point is often unstated or unclear.

-- In over 60% of sets of small-scale plans, it's unclear whether some dimensions are to partition wall finishes, wall framing, or centerlines.

-- Fifty percent of exterior wall details lack specifics on flashing and other waterproofing. The words "caulking" and "sealants" in notes are used interchangeably, as are notes of waterproofing," "moisture barrier," and "vapor barrier."

-- Construction joint locations, sizes, and types are missing, unclear, or contradictory in over 70% of the sets. The phrases, "expansion joints," "movement joints," "contraction joints," and "construction joints" are used interchangeably. Joint notation, besides being inaccurate, is frequently overdone and either duplicates or contradicts the specifications.

-- Thirty percent of relevant details fail to clearly show or note anchors, supports, backing, and fittings for mounted fix tures and equipment.

-- Offsets in exterior walls that have openings or other special construction are missing in 20% of exterior elevation drawing sheets.

-- Reference notes, such as "see Structural Drawings," "see Specifications," etc. appear throughout 80% of all working drawing sets. They're misleading nearly half the time, with the reference being nonexistent, very difficult to find, or contradictory to data on the reference drawing.

-- Up to 40% of the details in some drawing sets are not referenced from small-scale drawings. The details are there, but there's no way to know if they should be. This often requires explanatory phone calls and memos during construction.

-- Detail referencing is so bad in some large sets of documents that it can take up to 30 minutes to simply find a detail. Obviously, no one is willing to search for so long a time, and if it happens once, details won't be searched for again.

-- Slightly over 33% of all details serve no purpose. They're not needed and should not have been drawn at all.

In summary:

Fifteen percent of all sets of working drawings from licensed professionals are totally inadequate for bidding and construction.

Between 30% to 40% of all working drawing sheets have many lapses in important data.

ERRORS & OMISSIONS IN WORKING DRAWINGS continued

Only about 30% of working drawing sets can be considered fully professional.

If your office needs improvement, it may not take major house-cleaning to do it. Production managers tell us they get great results from very simple actions: Sit down with your drafters for an hour and review the items in this checklist and what they mean. Most drafters have never had such a learning session. Just one such meeting can eliminate most repetitions of the same problems in your office for years to come.

This is advice we published several years ago. Now if you'll add to the list of "red flags" the detailed lists that follow, your drafters will be among the best informed in the profession.

Notes:

CONTRACTORS WHO TURN ERRORS INTO PROFIT CENTERS

There's a new breed of contractor today, one who has been to business school and who may understand the common lapses in design practice better than most architects do.

Such contractors no longer worry about the protective boilerplate in contracts A/E's write. They know that much of what is written in contracts and specifications to protect A/E's from their mistakes won't hold up in arbitration or in court.

They don't worry about errors, omissions, and contradictions in the documents. They expect such problems and know where they're most likely to occur.

They don't worry about design changes or last-minute "clarifi-cations." They know how to predict such events and how to have their research and paperwork ready in advance of need. It's just more money in the bank.

Their main tool: well-planned management and preparation for extras and change orders.

Extras and change orders used to be as much a headache for contractors as for everyone else. Now contractors use systematic planning and management of extras to turn a negative into a positive . . . positive for them at least.

"It's a matter of attitude and thinking ahead," says the owner of a San Francisco general contracting firm. *"If you plan on changes and extras, you can list what's most likely to come up, and do a lot of the paperwork right after your bid is accepted. Then you're all set up to send your letters and claims without wasting a minute during the job."*

ERRORS & OMISSIONS IN WORKING DRAWINGS continued

CONTRACTORS WHO TURN ERRORS INTO PROFIT CENTERS continued

This contractor knows there will be errors, ambiguities and contradictions in most sets of bid documents. He knows where these most often occur. He knows how to spot drawings and specs that were done in a rush and/or haven't been carefully checked.

This contractor knows there will be errors, ambiguities and contradictions in most sets of bid documents. He knows where these most often occur. He knows how to spot drawings and specs that were done in a rush and/or haven't been carefully checked.

Here is what alert contractors look for in A/E documents:

-- Drawings without specific cross references. Notes such as "See Specs," "See Structural," "See details," etc. tell them the cross referencing wasn't carefully done or fully checked. The items referenced either won't be where they're supposed to be, or if they are, they're likely to be in conflict with other data.

-- Poor overall drawing organization -- details crowded into inappropriate sheets, crossed out and voided, items drawn at one scale and duplicated almost entirely elsewhere in the drawings at another scale. These are more indicators of minimal management of drawing production.

-- Inadequate detailing, such as construction details copied from manufacturers' catalogs that show products but don't show rough openings or connections to adjacent construction. Drawings with useless details and/or details that are grossly over-drawn, down to the screw threads, are considered a sure sign of inexperienced technical drafting and lax supervision.

ERRORS & OMISSIONS IN WORKING DRAWINGS continued

CONTRACTORS WHO TURN ERRORS INTO PROFIT CENTERS continued

-- Specifications with a mixture of writing styles and formats, contradictions, errors, and typos. These are clear clues that specs were hastily cut and pasted, without editing and correcting, to match the conditions of the job.

These are some of the first signs of potential added income for the contractor. After bidding, contractors use more detailed checklists to track down all likely errors and inconsistencies that will justify later claims.

There are more bonuses for the contractor if the A/E firm operates by day-to-day crisis management. So many change orders and extras may come up that A/E management will shoot from the hip and fire or transfer key staff members. Then those who follow will know little about the project and be less capable of challenging contractor claims. They'll just chalk the problems off to the bad guys who were fired, and the contractor will earn more with less resistance.

Another source of change-order income to contractors: delays and extra construction time, due to problems in the building design or the documents.

Recent court rulings have allowed contractors to sue A/E's, as well as building owners, for the economic loss of delays. An Illinois court has allowed two grounds for such suits: intentional misrepresentation and negligent misrepresentation. This could be applied 100% to negligent production drawings. Attorney Milton Lunch, writing in *Design Firm Man-agement & Administration Report*, quotes a ruling:

"One who, in the course of his business, profession . . . supplies false information for the guidance of others in their business transactions, is subject to liability for pecuniary loss

caused to them by their justifiable reliance on the information, if he fails to exercise reasonable care or competence in the obtaining or communicating of the information."

The double-whammy of lawsuit options where suits were not possible before adds muscle to the contractor's negotiating posture, as far as extras are concerned.

If these claims are premeditated, doesn't that go against the contract? Aren't bidders required to point out discrepancies and contradictions in the documents?

They're certainly premeditated by some contractors, and they've told us that it's good management practice to do so.

As for reporting problems in the documents, a contractor can't be held responsible for doing a complete, red-marked checkset of working drawings and specs. Most specs only call for contractors to notify the designer of glaring problems, and that's the most that can be expected.

THE CONTRACTOR'S "DISCOVERY" CHECKLIST

General Conditions usually require the contractor to call any discrepancies in documents and problems on the jobsite to the designer's attention. This instruction generally proves to be unenforceable.

A contractor may see signs of discrepancies and call a few obvious items to the designer's attention. But he or she is not expected to find them all; that's not an enforceable responsibility.

So in searching for potential change orders and extra payments on a project to be bid, a contractor will mainly seek the specific areas where discrepancies, contradictions, omissions, etc. are **likely** to occur and estimate extra work and costs accordingly.

Admittedly, this is hair splitting, but that's how it works.

What follows is an extensive list of problems contractors are trained to seek out, to turn inadequate design and documentation into a profit center.

The list was inspired by Mr. Civitello's book (for real detail, you should buy the book, by the way). We've also drawn on similar lists used by contractors, professional inspectors, and drawing checkers. It's all good information that will hopefully save you from some serious headaches, embarrassments, litigation, and money losses.

SITE-RELATED CAUSES OF DELAYS AND EXTRA COSTS

Contractors are instructed to review the following items in the project site plan, the survey, county property description, site inspection, and any related research deemed potentially useful. We've known of cases where the project designer never visited the site, thus giving the contractor a rather significant advantage in later disputes.

HIDDEN DATA IN SITE SURVEYS AND SOIL TESTS

Site data included in bid documents may be inadequate, hastily assembled, or otherwise unreliable.

Contractors are searching for:

__ Soil test data from an unknown or potentially unreliable source.

__ Inconsistent, subsurface boring depths.

__ Test borings spaced unevenly, outside the building line, and/or not as shown on the boring plan.

__ Outdated survey.

__ Survey inaccurate in major points.

__ Outdated borings and test results.

__ Soil test results affected by a season of testing.

__ Locations of existing manholes, drains and/or cisterns not as shown on the survey or drawings as existing conditions.

__ Adjacent wetlands.

__ High-tide or flood potential.

CIVIL ENGINEERING/SITE PLAN DRAWINGS

Contractors are searching for:

__ Full sheets of standard civil engineering drawings with few or no drawings of conditions unique to this site or this project.

__ Unusual foundation pier spacings.

__ Clumsy, hasty, or "low-budget" drafting of civil engineering drawings.

__ Existing and new grades not clearly indicated.

__ Cut-and-fill diagrams poorly drawn.

__ Drainage slopes that are minimal, ambiguous, or misdirected.

Notes:

____

____

____

____

____

____

FOUNDATION AND FRAMING DRAWINGS

Structural engineers are sometimes hired to stamp drawings done by unlicensed nonprofessionals or by developer/ owners themselves. They are extremely hazardous.

Contractors are searching for:

__ Full sheets of standard structural engineering drawings with few or no drawings of unique job conditions.

__ Structural details with no relationship to the project.

__ Structural-member detail drawings are all standards, handbook, or building code boilerplate.

__ Clumsy, hasty or "low budget" drafting of civil engineering drawings.

__ Extra-long spans.

__ Unusual column spacings.

__ Minimally-sized members throughout, without redundancy or safety factors.

__ Rebars not shown or clearly noted.

__ Structural connections not shown or clearly noted.

ARCHITECTURAL DRAWINGS

This is where most omissions, errors, and coordination lapses are likeliest to be found and where contractors focus their primary attention in the search for potential extras.

Contractors are searching for:

__ Drawing sets that are unusually bulky for a project of this type and size.

__ Unusually large drawing sheets relative to the project size and scope of work.

__ Drawing sheets that are unusually small and don't include baseline data that would be normal for a project of this type and size.

__ Rushed changes (changes shown on some drawings but not followed through in others).

__ Large numbers of "cloud" bubble notes showing late-date changes, deletions, and substitutions.

__ A noticeable number of "minor" errors such as the wrong scale for drawing title, misspellings, handwritten notes not erased from the final tracing before printing.

__ Plan match lines that are included but could have been avoided.

__ Plan match lines that are inconsistent on different plans.

__ Key plans that don't show locations of plan match lines.

__ North arrows that are inconsistent.

__ Out-of-sequence drawing sheets or erratic drawing numbering, indicating last-minute sheet deletions or insertions.

ELEVATORS AND STAIRS

Contractors are searching for:

__ Stair and elevator shaft drawings drawn at excessive size (evidence of design by junior personnel).

__ Small- and large-scale stair plans drawn by different personnel.

__ Ambiguous structural drawings, possible interferences with shaft.

__ Roof access not shown.

TOILET ROOM DRAWINGS

Contractors are searching for:

__ Toilet room drawings drawn at excessive size, evidence of design by junior personnel.

__ Small- and large-scale toilet room plans drawn by different personnel.

__ Possible inconsistencies with plumbing drawings.

__ Plumbing chase access not shown.

DRAWING SCALE AND SIZES

Contractors are searching for:

__ Unusually large drawing scale sizes of plans, elevations, etc., such as 1/2" plan and elevation scales, or 1/4" for larger building elevations and cross sections.

__ Unusually small, hard to read drawing scale sizes of plans, elevations, etc.

__ An unusually large number of elements labeled N.T.S. (not to scale).

NOTATION

Contractors are searching for:

__ Small, hard-to-read notes.

__ Notes indefinite as to references: "See Details," "See Structural drawings," "See Specifications."

__ Incomplete, inaccurate, or nonexistent reference notes.

__ Notes not professionally written (e.g., many undecipherable abbreviations).

__ Notes partially changed or erased, with remaining fragments.

__ N.I.C. work indicated, but unclear as to coordination and impact on contractor or subcontractors.

__ Lengthy notes (and notes that probably duplicate or contradict specifications).

ERRORS & OMISSIONS IN WORKING DRAWINGS continued

THE CONTRACTORS "DISCOVERY" CHECKLIST continued

DIMENSIONS

Contractors are searching for:

__ Dimensions that are too small to be easily read.

__ Dimensions that are incomplete.

__ Duplicate strings of dimensions.

__ Plans with interior and exterior dimensions of the same walls.

__ Elevation drawings that show horizontal dimensions that either duplicate or contradict the plans.

CROSS SECTIONS, EXTERIOR AND INTERIOR ELEVATIONS

Contractors are searching for:

__ Cross sections, floor heights, and ceiling and roof heights that don't coincide with exterior elevations.

__ Structural drawing framing heights that don't coincide with architectural drawings.

__ Excessive numbers of interior elevations, especially those repeated at larger scale, showing no significant construction.

__ Lack of clear indication of fixture mounting heights.

__ Duplication of detail key symbols, and door and window symbols shown on plan and also on elevations and sections.

CONSTRUCTION DETAILS

Contractors are searching for:

__ Details that don't clearly pertain to actual project construction.

__ Detail drawings that are relatively few for this type and size project.

__ When detail drawings are insufficient, no provision in the contract for procedures for ongoing clarification by design professional.

__ Detail drawings that differ widely in style, scale, and other ways (details have been borrowed from other sources and randomly pasted into this project's drawings).

__ Details excessive in quantity for the work to be performed.

__ Details excessive in size -- half- and full-size detail drawings that should really be done as shop drawings.

__ Excessive drawing of ready-made components or fabrications.

__ Inadequate drawing of actual connections of construction.

__ Lack of support backing for wall-mounted fixtures and handrails.

FINISH, DOOR, AND WINDOW SCHEDULES

Contractors are searching for:

__ Large, complex schedules, with numerous duplications.

__ Plan items not clearly linked to schedules.

__ Plans and elevations that both include the same door and window symbols -- a duplication of data that may lead to contradictions.

__ Plans and elevations that both include the same detail key symbols.

CEILING PLANS

Late, on-site discovery of conflicts in these aspects of work can be extremely time consuming and expensive.

Contractors are searching for:

__ Light fixture plans not coordinated with sprinkler head location plans.

__ Duct air supply and returns not coordinated with lighting and sprinkler plans.

__ Air supply and exhaust not located to avoid short circuiting heating and air conditioning.

__ Sprinkler head overlays showing conflicts with:
__ Lamps. __ Suspended ceiling channels.
__ Lighting tracks.

__ Smoke and fire detector locations showing conflicts with sprinklers, lamps, etc.

CEILING SPACES

Contractors are searching for:

__ Inadequate ceiling spaces.

__ Lack of clarity on who is responsible for coordinating trades.

__ Insufficient ceiling space for pipe slopes.

__ Piping and ductwork where beams or girders should be.

__ Large ducts that don't have open routes, to avoid beam interference.

__ Unknown height of specified light fixtures above ceilings.

__ Above-ceiling fire barriers breeched by large amounts of mechanical or electrical work.

__ Ceiling access panels not clearly detailed and specified.

__ Access panels improperly located and sized.

__ Inaccessible ceilings.

ROOF PLANS

Contractors are searching for:

__ Minimal drains in size and number.

__ Minimal roof slopes such that later settlement or framing shrinkage will cause ponding and deflection.

__ Drains and scuppers that empty onto walkways or public areas.

__ Slopes not clearly shown.

__ Plumbing, HVAC, and upper floor plans inconsistent with roof construction as shown.

EXISTING CONDITIONS

Contractors are searching for:

__ Inaccurate drawings of existing conditions.

__ New work not fully integrated with existing work to remain:
  __ In grading __ Exterior paving __ Floor elevations
  __ Exterior elevation alignments

__ Site showing traces of existing foundations that may have been covered over.

__ Site showing traces of basement areas or dump pits that may have been filled in.

__ Previous uses of property that may have involved toxic wastes.

__ History of archaeological research activity in the area.

CHASES AND PLUMBING WALLS

Contractors are searching for:

__ Inadequate chase sizes, especially for double-loaded walls.

__ Improperly sited and located cleanouts and maintenance panels.

__ Access panels not detailed and specified.

MECHANICAL ENGINEERING DRAWINGS

Contractors are searching for:

__ Sprinkler water mains that conflict with beams, ductwork, or electrical work.

__ Actual sizes of ductwork not clear.

__ Equipment hanger and support systems not indicated or detailed.

__ Lack of evidence of potential spatial conflicts in mechanical equipment.

__ Insufficient room for mechanical equipment.

__ Inadequate maneuvering space to move large mechanical equipment into place.

__ Non-centralized mechanical spaces and main ducts.

__ "Spaghetti runs" of ductwork.

__ Basement wall openings for utilities, pipes and drainage uncoordinated with foundation or structural drawings.

CONSULTANT DRAWING CHECKING AND COORDINATION

This checklist provides a useful supplement to the contractor's "discovery" checklist.

COMMON ARCHITECTURAL DRAWING ERRORS AND COORDINATION PROBLEMS

__ Existing work, work to remain or to remove, and new work unclearly identified and differentiated.

__ Exterior elevations that don't match doors, windows, roof lines, and expansion joints on the plans.

__ Building cross sections that lag behind and are uncoordinated with plans and elevations.

__ Rough openings for doors and windows that are too large or too small (especially in masonry).

__ Expansion joints that aren't continuous throughout the building.

__ Room wall/floor/ceiling construction that doesn't match the finish schedule.

__ Door, window, and frame schedules that don't reflect changes in doors and windows on the plans.

CIVIL ENGINEERING COORDINATION

__ Site plans for interferences of underground utilities
  __ allow for:
  __ Power __ Telephone __ Water __ Sewer __ Gas
  __ Storm drainage __ Fuel lines __ Grease traps
  __ Fuel tanks

__ Site plans for interferences between new drives, sidewalks, and other new sitework
  __ allow for:
    __ Telephone poles __ Pole guy wires
    __ Street signs __ Drainage inlets
    __ Valve boxes __ Manholes

__ Civil earthwork grading and excavation plans
  __ coordinate with architectural and landscape plans for:
    __ Clearing __ Grading
    __ Sodding, grass, and mulch
    __ Other landscaping

__ Civil drawings with fire hydrant and street light pole locations
  __ coordinate with other drawings:

    __ Architectural Sitework __ Electrical
    __ Plumbing

__ Profile sheets show underground utilities:
  __ Conflicts are avoided in elevation as well as plan.

__ Plan and profile sheets of drainage systems and manholes
  __ Scaled dimensions match with written dimensions:
    __ In plans __ In profiles

CONSULTANT DRAWING CHECKING AND COORDINATION
continued

__ Final finish grade and pavement elevations relative to manholes and valve boxes:

__ Tops of manholes/utility boxes are flush with finish grade, pavement, walks, streets:
__ Sewer __ Power __ Telephone __ Drains

__ All existing and final grades are noted at every point of change:

__ Dimensions are reasonable, without widely varying differences.

__ Grades allow for run-off drainage without danger of ponding.

Notes:

____

____

____

____

____

____

____

____

____

STRUCTURAL DRAWINGS COORDINATION

__ Overlay and compare:
- __ Column lines on structural and architectural.
- __ Column locations on structural and architectural.
- __ Perimeter slab on structural and architectural.
- __ Depressed and raised slabs on structural and architectural.
- __ Slab elevations on structural and architectural.

__ Structural drawings:
- __ Foundation piers are identified.
- __ Foundation beams are identified.

__ Roof framing plan column lines and columns line up with lower levels.

__ Foundation plan column lines and columns line up with upper levels.

__ Perimeter roof line matches Architectural roof plan.

__ All columns and beams are identified and listed in column and beam schedules.

__ Column lengths are all shown in column schedule.

__ All sections are properly labeled.

__ All expansion joint locations match Architectural.

__ Dimensions agree with Architectural.

__ Drawing notes agree with Specifications.

MECHANICAL AND PLUMBING COORDINATION

__ All new electrical, gas, water, sewer, etc. lines connect to existing.

__ All plumbing fixture locations are coordinated with architectural.

__ Plumbing fixtures are coordinated with fixture schedule and/or specs.

__ Storm drain system aligns with drains and leaders shown on roof plan.

__ Pipes are sized and all drains are connected and do not interfere with foundations.

__ Wall chases are provided on architectural, to conceal vertical piping.

__ Sanitary drain system pipes are sized, and all fixtures are connected.

__ HVAC floor plans match architectural.

__ Sprinkler heads are shown in all rooms.

__ All sections are identical to architectural/structural.

__ Adequate ceiling height exists at worst-case duct intersections.

__ All structural supports required for mechanical equipment are indicated on structural drawings.

CONSULTANT DRAWING CHECKING AND COORDINATION
continued

__ Dampers are indicated at smoke and fire walls.

__ Diffusers are coordinated with architectural reflected ceiling plan.

__ All roof penetrations (ducts, fans, etc.) are indicated on roof plans.

__ All ductwork is sized.

__ All air conditioning units, heaters, and exhaust fans match architectural roof plans and mechanical schedules.

__ Mechanical equipment will fit in spaces allocated.

Notes:

____

____

____

____

____

____

____

____

ELECTRICAL COORDINATION

__ All plans match the architectural.

__ Light fixtures match the architectural reflected ceiling plan.

__ All powered equipment has electrical connections.

__ All panel boards are properly located and are shown on the electrical riser diagram.

__ There is sufficient space for all electrical panels to fit.

__ Electrical panels are not recessed in fire walls.

__ Electrical equipment locations are coordinated with site paving and grading.

__ Motorized equipment:

  __ Equipment shown is coordinated with electrical drawings.

  __ Horsepower ratings are verified.

  __ Voltage requirements verified.

Notes:

____

____

CEILING PLAN COORDINATION

__ Reflected ceiling plan agrees with room plans.

__ Ceiling materials match finish schedule.

__ Light fixture layout matches Electrical.

__ Ceiling diffusers/registers match Mechanical, including all soffits and vent locations.

KITCHEN PLAN COORDINATION

__ Equipment layout agrees with architectural plans.

__ All equipment is connected to utility systems.

Notes:

____

____

____

____

____

____

____

____

____

UNIFORM DRAWING PROCEDURES AND CONTENT

Drafters often need explicit and detailed step-by-step instructions regarding content and procedures for completing working drawing sheets. This is especially important in CADD where graphics are often assembled rather than drawn, and as many as 20 or more files may comprise a single drawing sheet.

It's easy, in the intricacies of CADD production, to lose sight of the purpose of the drawings, their required content, and the most efficient steps for creating them.

Thus this checklist, the beginning of what may become a nationally accepted procedural standard for production and quality control.

UNIFORM DRAWING PROCEDURES -- SITE PLANS

THE OBJECTIVE is to create architectural site plan information that will:

1) Guide engineering consultants -- civil, structural, and plumbing -- who will be designing earthwork, paving, and drainage according to architectural design intentions.

2) Provide information to guide other engineers who may need site information, such as for electrical hook-up, water, and other utilities.

3) Convey information about the site sufficient to satisfy regulatory agencies such as the building and zoning departments. Much of this work will be completed during the Design Development phase.

4) Show the boundaries of construction to the contractors, and inform the contractors of existing conditions and new work in sufficient detail that they can create reasonably accurate estimates of construction costs and provide competitive bids for the work. In some cases we may work with a contractor or owner-builder who will do the work and who is well acquainted with the site. In such cases, you will receive special instructions as to the content and comprehensiveness of the drawings.

SITE PLANS -- CONTENT CHECKLIST

BASELINE CONTENT

Baseline content is that which is required for building department approval and construction either by an owner-builder or through a negotiated contract with an owner-selected contractor.

__ Property outline only and/or minimum data as specified by building department

__ Easements and rights of way

__ Points of utility connection

__ Indication of property slopes and drainage

__ Storm drains, new and existing

__ Building outline or footprint with exterior wall dimensions to property boundary lines

__ Building overhang dimensions

__ Street access, driveway dimensions

__ Parking space as required

__ Handicap parking, pavement markings, signs, ramps and access as required

__ Utility meters and hook-ups

__ Print of the survey may be bound into the drawings for reference

__ General notes regarding work not shown in detail

STANDARD CONTENT

Standard content is that typically required for building department approval, construction cost estimating and bidding, and complete construction by a qualified general contractor.

Includes previously listed content plus:

__ Spot check of accuracy of surveyor's drawings and notes of discrepancies

__ Zoning check and confirmed setback limit dimensioned

__ Site reference photos to establish views, neighbor proximity, etc.

__ Existing and new site contours

__ Existing and new finish grades at all corners of building

__ Grade slopes at building line

__ Dimensional allowance for final grading elevations with addition of topsoil

__ Building overhang dimensions

__ Roof plan sometimes shown on site

__ Site drain slopes for all drainage, including roof

__ Driveway and street centerline elevations and side elevations

__ Pavement construction joints and movement joints with detail and specification references

__ Existing and new curb inlets, catch basins, manholes, flumes and spillways

UNIFORM DRAWING PROCEDURES AND CONTENT continued

SITE PLANS -- CONTENT CHECKLIST continued

__ Pavement slopes to drain

__ Existing site elements to remain/remove

__ Cut and fill grading profiles

__ Utility meters and hook-ups

__ Underground fuel or other storage

__ Buried cables and main warning signs

__ Soil test reference data

__ Drawings for contractor, to assist with construction and temporary facilities planning

__ Retaining walls, fences and gates

__ Signs, kiosks, and other site appurtenances

Notes:

____

____

____

____

____

____

____

EXTENDED CONTENT

Extended content is that required for projects requiring extra design office attention because of complex construction, elaborate detailing, and/or extended collaboration with specialized consultants.

Includes previously listed content plus:

__ Confirmed site dimensions from on-site review and measurements at site

__ Meets and bounds confirmation

__ Aerial photographs and conversion of aerial photos and survey to same scale for overlay comparison

__ Site photo and/or video survey encompassing all major site features and surroundings

__ Separate screened prints of background site work to show:
- __ Grading and building excavation
- __ Drainage excavation
- __ Electrical excavation and overhead work
- __ Site lighting and electrical outlets
- __ Plumbing excavation: __ Gas __ Sewer __ Water
- __ HVAC excavation
- __ Landscaping
- __ Site furniture and appurtenances

__ Construction fence design and plan

__ Temporary roads, parking, storage, other construction facilities

__ Roof drainage and final resolution of drainage

__ Landscaping plan:
- __ Plant selection
- __ Plant purchase and supervised installation
- __ Irrigation and/or sprinkler system planning

__ Existing trees and landscaping protection plan and specifications

__ Comprehensive "as-built" record drawings of final site work or review and OK of contractors' record drawings

__ Additional plans as required:
- __ Temporary construction facilities
- __ Percolation test plan
- __ Soils testing boring schedule and profile
- __ Test pit and boring plan
- __ Test boring locations
- __ Copy of test profile from geotechnical engineer
- __ Temporary erosion control
- __ Grading plan
- __ Demolition plan
- __ Excavation plan
- __ Drainage

__ Photos of major site features -- existing to remain, modify, to remove and store, demolish

Notes:

____

____

____

____

____

____

____

CHECKLIST OF SITE PLAN DRAWINGS

Site plans may incorporate individual or combined sheets, including the following:

__ SURVEY

__ STAKING PLAN

__ TEST PIT AND BORING PLAN

__ GRADING PLAN

__ DEMOLITION PLAN

__ EXCAVATION PLAN

__ CONSTRUCTION WORK AND TEMPORARY FACILITIES

__ ARCHITECTURAL SITE PLAN
- __ general construction
- __ paving, walkways and parking
- __ site furniture and appurtenances

__ DRAINAGE PLAN

__ LANDSCAPING PLAN

__ ELECTRICAL
- __ supply
- __ outdoor lighting

__ HVAC

__ PLUMBING
- __ supply
- __ irrigation
- __ sprinkler

SITE PLANS -- GENERAL REFERENCE DATA

__ DRAWING TITLE AND SCALE

__ ARROWS SHOWING COMPASS NORTH AND REFERENCE NORTH

__ SMALL-SCALE LOCATION OR VICINITY MAP SHOWING NEIGHBORING STREETS AND NEAREST MAJOR HIGHWAY ACCESS
(Can be copy of a portion of the local road map)

__ NOTE REQUIRING BIDDERS TO VISIT SITE AND VERIFY CONDITIONS BEFORE SUBMITTING BIDS

__ WORK NOT IN CONTRACT

__ PROPERTY SIZE IN SQUARE FEET OR ACRES

__ CONTRACT LIMIT LINES

__ LOCATION AND DIMENSIONS OF CONSTRUCTION

__ DETAIL KEYS

__ DRAWING CROSS REFERENCES

__ SPECIFICATION REFERENCES

__ LEGAL SETBACK LINES

__ EASEMENTS

__ SITE PHOTOS
(Usually printed adjacent to the site plan drawing with connecting arrow lead lines to show the exact areas of site represented by photos.)

__ LEGENDS OF SITE PLAN SYMBOLS AND MATERIAL INDICATIONS

__ LANDSCAPE CONSULTANT
__ address __ phone number

SITE PLANS -- CONTENT CHECKLIST continued

SITE PLANS -- GENERAL REFERENCE DATA continued

__ SURVEYOR
 __ address and phone number __ registration number

__ CIVIL ENGINEER
 __ address and phone number __ registration number

__ SOILS ENGINEER
 __ address and phone number __ registration number

__ SOIL TEST LAB
 __ address and phone number
 __ registration number

__ TEST BORING CONTRACTOR
 __ address and phone number

__ DESCRIPTION OF SOIL TYPE AND BEARING

__ PERCOLATION TEST PLAN

__ SOILS TESTING BORING SCHEDULE AND PROFILE

__ TEST BORING LOCATIONS

__ BORING TEST PROFILE FROM SOILS ENGINEER
 (May be separate drawing.)

Notes:

____

____

____

____

____

FLOOR PLANS

THE OBJECTIVE is to create architectural floor plan information that will:

1) Guide engineering consultants -- structural, HVAC/plumbing, and electrical -- who will be designing framing, mechanical systems, and lighting and electrical systems, according to design intentions.

2) Provide information to guide other consultants who may need plan information, such as interior designers, kitchen planners, acoustical consultants, etc.

3) Convey information about the building sufficient to satisfy regulatory agencies such as the building department and zoning agencies. Give particular attention to fire safety -- exits, exit corridors, sprinklers, isolation of hazardous equipment, etc. Fire exit planning should be completed and approved during the Design Development phase. Give similar consideration to handicap accessibility.

4) Show the limits of construction to the contractors, and inform the contractors of existing conditions and new work in sufficient detail that they can create reasonably accurate estimates of construction costs and provide competitive bids for the work. In some cases we may work with a contractor or owner-builder who will do the work and who is well acquainted with the site. In such cases, you will receive special instructions as to the content and comprehensiveness of the drawings.

5) Show special construction circumstances, such as work or items to be provided and/or installed by the Owner, existing work to remain, be repaired or removed, and phased construction and future additions.

FLOOR PLANS -- CONTENT CHECKLIST

BASELINE CONTENT

Baseline content is that which is required for building department approval and construction, either by an owner-builder or through a negotiated contract with an owner-selected contractor.

__ Rooms dimensioned with room name -- interior finish to finish size, such as "Dining 12' x 14'"

__ Walls and partitions with centerline or surface dimensions

__ Walls and partitions with keyed materials indications

__ Doors and windows with sizes/types noted at each unit, no separate door or window schedules

__ Ceiling or roof framing shown by arrows, to indicate direc tion of frame, with note of member sizes and spacings

__ Stairs:
  __ Arrow down or up, with noted number of treads
  __ Finish clear opening dimensions

Notes:

____

____

____

____

STANDARD CONTENT

Standard content is that typically required for building department approval, construction cost estimating and bidding, and complete construction by a qualified general contractor.

Includes previously listed content, plus:

__ Dotted-line indications for openings, soffits, etc. at ceiling with notes, dimensions, and detail references

__ New work in solid line in contrast to screened background of existing work to remain

__ Existing work to remain identified with symbols or patterns

__ Door and window symbols keyed to door and window schedules

__ North arrow and reference north

__ Detail keys at junctures of different floor materials

__ Detail or schedule keys at floor saddles and thresholds

__ Door and window symbols keyed to door and window frame or rough opening schedules

__ Door and window frame schedules keyed to details

__ All door and window details included on door/window frame schedule sheets

__ Notes of depressed slabs or framing to accommodate divergent thicknesses of finish floors

__ Sound isolation walls

__ Cabinet work keyed to cabinet schedule

FLOOR PLANS -- CONTENT CHECKLIST continued

__ Stairs:
  __ Rough and clear opening dimensions
  __ Stair profile section and stair detail reference keys
  __ Noted non-slip treads and other safety elements

__ Dimensions of chases, shafts, and furred walls

__ Fixtures such as drinking fountains, fire hose cabinets, etc., keyed to fixture schedule and details

Notes:

____

____

____

____

____

____

____

____

____

____

____

____

EXTENDED CONTENT

Extended content is that required for projects requiring extra design office attention because of complex construction, elaborate detailing, and/or extended collaboration with specialized consultants.

Includes previously listed content, plus:

__ North arrow, reference north, and symbols at exterior walls, keyed to Exterior Elevation drawings

__ Finish flooring patterns or reference to enlarged pattern drawings

__ Data references to show relative floor heights of balconies, landings, mezzanines, etc.

__ Expansion space notes or detail references for perimeters of expansive flooring

__ Numbered and lettered structural grid coordinates with primary framing dimensions

__ Framing plan for larger projects as part of structural drawings

__ Walls and partitions with dimensions to framing and substructure, plus cumulative dimensions across plan

__ Walls and partitions with keyed materials indications linked to materials schedule

__ Simplified finish schedules included on floor plan sheets

__ Keys or symbols to break-out work of different trades or contracts

UNIFORM DRAWING PROCEDURES AND CONTENT continued

FLOOR PLANS -- CONTENT CHECKLIST continued

__ Detail keys at walls and floor anchors for mounted equipment and fixtures

__ Interior wall construction keyed to enlarged wall plan, and vertical sections showing acoustical and fire-rated treatments

__ Exterior wall construction keyed to enlarged wall plan and vertical sections, showing framing, detail keys, thermal insulation, and waterproofing

__ Floor and ceiling plans keyed to floor and ceiling detail sections showing framing, detail keys, acoustic and fire-rated treatments

__ Noted and detailed vibration and noise control at equipment supports

__ Stairs:

- __ Treads and risers with noted T & R dimensions
- __ Rough and clear opening dimensions
- __ Stair section and stair detail reference keys
- __ Noted non-slip treads and other safety elements

Notes:

___

___

___

___

FLOOR PLANS & INTERIOR ELEVATIONS GENERAL REFERENCE DATA

__ DRAWING TITLES AND SCALE

__ ARROWS SHOWING COMPASS NORTH AND REFERENCE NORTH

__ MODULAR GRID OR STRUCTURAL COLUMN GRID WITH NUMBER AND LETTER COORDINATES

__ SQUARE FOOTAGE TOTALS:
- __ building
- __ auxiliary structures
- __ decks and balconies
- __ rooms (name or number)
- __ exterior areas

__ KEY PLAN

__ INTERIOR ELEVATION ARROW SYMBOLS AND REFERENCE NUMBERS

__ EXTERIOR ELEVATION ARROW SYMBOLS AND REFERENCE NUMBERS

__ OVERALL CROSS SECTION LINES AND KEYS

__ MATCH UP LINE, OVERLAP LINE, AND REFERENCE IF FLOOR PLAN IS CONTINUED ON ANOTHER SHEET

__ SPACE FOR ITEMS N.I.C.

__ NOTE ON N.I.C. ITEMS TO BE INSTALLED, CONNECTED BY CONTRACTOR

__ OUTLINE OF FUTURE BUILDING ADDITIONS

__ REMODELING
- __ existing work to remain as is
- __ existing work to relocate
- __ existing work to be removed
- __ existing work to be repaired or altered

UNIFORM DRAWING PROCEDURES AND CONTENT continued

FLOOR PLANS & INTERIOR ELEVATIONS

__ GENERAL NOTES

__ BUILDING CODE/FIRE CODE REFERENCES

__ MATERIALS HATCHING

__ DETAIL KEYS

__ DRAWING CROSS REFERENCES

__ SPECIFICATION REFERENCES

Notes:

EXTERIOR ELEVATIONS

THE OBJECTIVE is to create exterior elevation information that will:

1) Guide consultants who may be concerned about exterior features such as exterior structural members, location of air intakes and exhausts, and special wall- or roof-mounted electrical or mechanical equipment.

2) Show heights of windows, doors and other openings in exterior walls. Show heights of structural members such as bearing plates, floor-to-floor heights, parapet heights, etc. Such information is mainly for the framing contractor. Door and window symbols should only be on floor plans. Specific types and sizes will be conveyed in Schedules.

3) Show generic materials for walls and roof. Show roof slopes and major roof features.

4) Show roof drain leaders and grade level or below-grade drains.

5) Show relationship of foundations and exterior walls to existing grade, finish grade, and grade that may be added for landscaping. Unless handled by special drainage, all grades must slope away from the building.

6) Convey information about the building sufficient to satisfy regulatory agencies such as the building department and zoning agencies. The relationship of buildings in close proximity to this building and this building's relationship to the lot line are shown in the site plan, but some special features, such as roof overhangs, balconies, etc. may have legal or zoning implications that need to be documented.

UNIFORM DRAWING PROCEDURES AND CONTENT continued

EXTERIOR ELEVATIONS continued

7) Show heights of construction to the contractors. Inform them of existing conditions and new work in sufficient detail that they can create reasonably accurate estimates of construction costs and provide competitive bids for the work. In some cases we may work with a contractor or owner-builder who will do the work and who is well acquainted with the site. In such cases, you will receive special instructions as to the content and comprehensiveness of the drawings.

8) Show special construction circumstances, such as work or items to be provided and/or installed by the Owner, existing work to remain, be repaired or removed, and phased construction and future additions.

Notes:

____

____

____

____

____

____

____

____

____

EXTERIOR ELEVATIONS CONTENT

BASELINE CONTENT

Baseline content consists of that which is required for building department approval and construction either by an owner-builder or through a negotiated contract with an owner-selected contractor.

__ Exterior elevations of primary sides of the building, labeled as shown on key plan

__ Exterior material notes or pattern indications

__ Doors and windows with general height dimension for door and window finish openings

__ Overhangs

__ Wall and overhang dimensions to property lines, if close to setback limits

__ Finish grades

__ Below-grade construction

__ Retaining walls, paving, planters, curbs, etc. adjacent to structure

__ Floor, ceiling, roof heights

__ Roof slopes

__ Major openings

STANDARD CONTENT

Standard content is that typically required for building department approval, construction cost estimating and bidding, and complete construction by a qualified general contractor.

Includes previously listed content, plus:

__ Notes whether vertical dimensions are to finish surfaces, subfloor, or framing

__ Notes whether opening dimensions are rough or finish openings

__ Air intakes coordinated with HVAC drawings (confirmed as safely away from building or car exhausts)

__ Adjacent fences and yard walls

__ Adjacent site furniture such as decks, pavement, railings

__ Adjacent major landscaping features that might affect construction

__ Site features to remain, features to remove

__ Wall-mounted exterior lights and alarms

__ Flagpoles, signs, plaques, other wall-mounted appurtenances

__ Hose bibbs, hydrants, standpipes

__ Downspouts, rain leaders, roof drains and scuppers

__ Roof gutter slopes

__ Catch basins, splash blocks, trench drains for roof drainage

UNIFORM DRAWING PROCEDURES AND CONTENT continued

EXTERIOR ELEVATIONS CONTENT continued

__ Wall-mounted marquees, awnings, canopies, etc.

__ Wall-mounted appurtenances which are N.I.C., shown but drawn dotted line or clearly labeled N.I.C.

__ Overhead utility connection points

__ Demountable or movable roof equipment in dotted line with Identification Note

__ Roof railings and guards with dimension of required safety height

__ Vertical joints for expansion/contraction and specification reference note

__ Horizontal movement joints for frame shrinkage and specification reference note

__ Flashing and waterproofing indications with specification reference notes

Notes:

___

___

___

___

___

EXTENDED CONTENT

Extended content is that required for projects requiring extra design office attention because of complex construction, elaborate detailing, and/or extended collaboration with specialized consultants.

Includes previously listed content, plus:

__ Rough opening structure and substructure dimensions

__ Sill and lintel substructure heights

__ Vertical working dimensions from an established start point such as finish slab or foundation

__ Background screened image of overall building cross section with exterior wall construction in solid line

__ Coded dimension indications for rough, substructure, and finish dimensions

Notes:

____

____

____

____

____

UNIFORM DRAWING PROCEDURES AND CONTENT continued

EXTERIOR ELEVATIONS

GENERAL REFERENCE DATA

__ DRAWING TITLE AND SCALE

__ SHEET NUMBER AT TITLE BLOCK

__ TITLES OR SYMBOL KEYS TO IDENTIFY EACH ELEVATION VIEW

__ KEY PLAN SHOWING NORTH ARROW AND ELEVATION VIEW LOCATIONS

__ MODULAR GRID OR STRUCTURAL COLUMN GRID WITH NUMBER AND LETTER COORDINATES

__ PROPERTY LINES

__ SETBACK LINES

__ OUTLINE OF ADJACENT STRUCTURES

__ OUTLINE OF FUTURE BUILDING ADDITIONS

__ MATCH UP LINE, OVERLAP LINE, AND REFERENCE IF DRAWING IS CONTINUED ON ANOTHER SHEET

__ ITEMS N.I.C., N.I.C. ITEMS INSTALLED OR CONNECTED BY CONTRACTOR

__ REMODELING
- __ work to be removed
- __ work to be repaired
- __ work to remain
- __ work to be relocated

__ GENERAL NOTES

__ BUILDING CODE REFERENCES

__ DRAWING CROSS REFERENCES

__ SPECIFICATION REFERENCES

__ CROSS SECTION LINES AND KEYS

__ WALL SECTION LINES AND KEYS

__ MATERIALS HATCHING
(Use only partial materials indications on building area.)

__ DETAIL KEYS
(Not for doors, windows or storefront details. Those details are to be referenced from schedules, not elevations or plans.)

ROOF PLANS

THE OBJECTIVE is to create roof plan construction information that will:

1) Guide engineers who may be concerned about exterior features such as roof slopes, overhangs, structural members, location of air intakes, vents, exhausts, roof-mounted electrical or plumbing equipment.

2) Show roof drains, gutters, and scuppers with ample clarity as to placement and sizes. Any drainage problems or potential conflicts with other construction should be resolved in the drawings, not left for on-site resolution.

3) Show skylights, monitors, shafts, elevator and stair penthouses, curbs, parapets, hips, and valleys, with detail keys. Assure that skylights, shafts, etc. are coordinated with the upper floor plan.

4) Show the generic materials of the roof. Be sure roof slopes, parapets, and other major roof features are coordinated with exterior elevations and cross sections.

5) Convey information about the building sufficient to satisfy regulatory agencies such as the building department and zoning agencies. The relationship of other buildings in close proximity to this building and this building's relationship to the lot line are shown in the site plan, but some special features, such as roof overhangs, balconies, etc. may have legal or zoning implications that need to be documented.

6) Show roof construction intent to the contractors, in sufficient detail that they can create reasonably accurate estimates of construction costs and provide competitive bids for the work. In some cases we may work with a contractor or owner-builder who will do the work and who is well acquainted with the site.

In such cases, you will receive special instructions as to the content and comprehensiveness of the drawings.

7) Show special construction circumstances, such as work or items to be provided and/or installed by the Owner, existing work to remain, be repaired or removed, and phased construction and future additions.

ROOF PLANS CONTENT

BASELINE CONTENT

Baseline content is that which is required for building department approval and construction either by an owner-builder or through a negotiated contract with an owner-selected contractor.

__ Roof surface with materials, overhangs and slopes shown in combination with Site Plan

__ Roof overhang edge dimensions to property lines and setback lines

__ Drain indications and flashing detail keys

Notes:

____

____

____

STANDARD CONTENT

Standard content is that typically required for building department approval, construction cost estimating and bidding, and complete construction by a qualified general contractor.

Includes previously listed content plus:

__ Roof construction plan in same scale as floor plans

__ Roof finish materials, overhangs, and slopes

__ Indication of direction of roof framing

__ Exhaust vents, fresh air intakes (confirmed that fresh air intakes are not near or in wind path of exhaust vents)

__ Scuppers, overflow scuppers and roof drainage slopes in inches per linear foot

__ Drain types, sizes, and detail keys

__ Dotted-line indications for openings, soffits, etc. at ceiling with notes, dimensions, and detail

__ Detail keys for eaves, parapets, drains, gutter guards, sky lights and roof-mounted equipment

__ Notation of roof insulation

Notes:

____

____

____

EXTENDED CONTENT

Extended content is that required for projects requiring extra design office attention because of complex construction, elaborate detailing, and/or extended collaboration with specialized consultants.

Includes previously listed content, plus:

__ Roof framing plans as separate sheets from plans showing finish roofing

__ Shadow print of upper floor plan in combination with solid roof plan, to show coordination of structural frame, drains, skylights, etc.

__ Shadow prints of upper floor roof plan with combination of solid line roof plan, to show coordination of roof with work of engineering disciplines:
- __ Plumbing
- __ HVAC
- __ Electrical

__ Roof plan and details referenced to specifications section numbers

__ Roof railing and parapet heights

__ Utility service entries at roof

__ Types and location of anchors, supports and guys for roof-mounted equipment

ROOF PLANS -- REFERENCE INFORMATION

__ BUILDING OUTLINE AND ROOF OUTLINE
(Broken-line building perimeter where shown below roof overhang.)

__ OVERALL EXTERIOR WALL DIMENSIONS

__ ROOF OVERHANG DIMENSIONS

__ OUTLINE OF FUTURE ADDITIONS

__ CONNECTION WITH EXISTING STRUCTURES

__ BUILDING AND ROOF EAVE DIMENSIONS

__ DIMENSIONS TO PROPERTY LINES

__ BUILDING LOCATION REFERENCED TO BENCH MARK

__ REQUIRED PROPERTY SETBACK LINES

__ ALL PROJECTIONS LABELED AND DIMENSIONED SHOWING COMPLIANCE WITH PROPERTY LINE AND SETBACK REQUIREMENTS

__ STAIR NUMBERS

__ OUTLINE OF OPENINGS AND PROJECTIONS
- __ bays
- __ areaways
- __ balconies
- __ marquees
- __ canopies
- __ landings
- __ decks
- __ steps
- __ sills

CROSS SECTIONS

THE OBJECTIVE is to create cross section information that will:

1) Guide structural and mechanical engineering consultants who need to know the sizes and height relationships of spaces, especially floor-to-ceiling and floor-to-floor heights.

2) Identify and provide detail keys for areas of special construction such as through-floor shafts, stair wells, elevator shafts, major structural members, and special roof-mounted electrical or mechanical equipment.

3) Show heights of openings in walls. Show heights of structural members such as girders and headers, floor-to-floor heights, parapet heights, etc. This information is mainly for the framing contractor.

4) Show all information with simple, uncluttered graphics. Avoid showing information that may be duplicated or contradicted on other drawings, such as "back walls," doors and windows, or door and window symbols.

5) Show generic materials for floors, walls, and roof, and show roof slopes and major roof features.

6) Show the relationship of foundations and exterior walls to existing grade, finish grade, and grade that may be added for landscaping. Unless handled by special drainage, all grades must slope away from the building.

7) Convey information about the building sufficient to satisfy regulatory agencies such as the building department and zoning agencies. The relationship of buildings in close proximity to this building and this building's relationship to the lot line

are shown in the site plan, but some special features may have legal or zoning implications that need to be documented.

8) Show special construction circumstances, such as work or items to be provided and/or installed by the Owner, existing work to remain, be repaired or removed, and phased construction and future additions.

9) Show heights of major construction elements to the contractors. Inform them of existing conditions and new work in sufficient detail that they can create reasonably accurate estimates of construction costs and provide competitive bids for the work.

Notes:

____

____

____

____

____

____

____

____

____

CROSS SECTIONS CONTENT CHECKLIST

BASELINE CONTENT

Baseline content is that which is required for building department approval and construction either by an owner-builder or through a negotiated contract with an owner-selected contractor.

__ Cross-sections and wall sections are not necessarily provided except to show exceptional construction

__ Standard wall construction sections may be photocopied from handbooks

__ Footings and foundation walls (unless covered in structural drawings)

__ Bearing walls including rough openings

__ Floor construction

__ Balconies and mezzanines

__ Parapets

__ Overhangs

__ Roof framing

Notes:

____

____

____

STANDARD CONTENT

Standard content is that typically required for building department approval, construction cost estimating and bidding, and complete construction by a qualified general contractor.

Includes previously listed content, plus:

__ Interior partitions

__ Suspended ceilings, where appropriate, to show relationship to structural framing or mechanical equipment

__ Mechanical equipment, where appropriate, to show relationship structural framing

__ Suspended heavy equipment in relationship to framing

__ Wall-mounted equipment in relationship to structural frame

__ Roof appurtenances, such as penthouse and stair bulkhead, in relationship to framing

Notes:

____

____

____

____

____

____

EXTENDED CONTENT

Extended content is that required for projects requiring extra design office attention because of complex construction, elaborate detailing, and/or extended collaboration with specialized consultants.

Includes previously listed content, plus:

__ Key plan with identification of cross section cut points

__ Special sections to show thru-building shafts and light shafts

__ Special structural systems, bracing, stiffeners, etc.

__ Directly connected blow-up details of special connections

Notes:

____

____

____

____

____

____

____

CROSS SECTIONS -- REFERENCE INFORMATION

__ KEY PLAN IDENTIFICATION OF CROSS SECTION CUT POINTS

__ EXISTING AND NEW FINISH GRADE LINES WITH ELEVATIONS

__ HEIGHTS AND ELEVATION POINTS
- __ finish floor to finish floor or subfloor to subfloor
- __ floor or subfloor to finish ceiling
- __ floor or subfloor to underside of beams, headers and lintels
- __ upper floor to roof substructure

__ NOTE WHETHER DIMENSIONS ARE TO FINISH SURFACES, ROUGH SURFACES, OR STRUCTURE

Notes:

____

____

____

____

____

____

____

____

____

WALL SECTIONS

THE OBJECTIVE is to create wall section information that will:

1) Guide structural consultants who need to know the sizes and relationships of major features, such as through-wall openings, headers, wall mounted and cantilevered construction, floor-to-ceiling and floor-to-floor heights.

2) Identify and provide detail keys for areas of special construction such as major structural members, wall-mounted equipment, cantilevers, parapets, etc.

3) Show heights of openings in walls. Show heights of structural members such as girders and headers, floor-to-floor heights, parapet heights, etc.

4) Show generic materials for floors, walls, ceilings, and roof, and show any intricacies of materials such as wall plates, brick or hollow masonry unit courses, etc.

5) Show relationship of foundations and exterior walls to existing grade and finish grade. Show drains and weep holes or key detail references for these elements.

6) Convey information about the building sufficient to satisfy regulatory agencies such as the building department.

7) Show heights of construction to the contractors, and inform them of existing conditions and new work in sufficient detail that they can create reasonably accurate estimates of construction costs and provide competitive bids for the work.

REFLECTED CEILING PLANS

THE OBJECTIVE is to create reflected ceiling plan information that will:

1) Guide our engineering consultants -- structural, HVAC /plumbing, and electrical -- who will be designing framing, mechanical systems, lighting and electrical systems according to our design intentions.

2) Resolve potential conflicts between above-ceiling construction -- lighting fixtures, structural members, ductwork, and piping -- before they become job-site problems.

3) Provide information to guide other consultants who may need plan information, such as lighting consultants, interior designers, kitchen planners, acoustical consultants, etc.

4) Convey information about the building sufficient to satisfy regulatory agencies such as the building department and zoning agencies. Give particular attention to fire safety -- sprinklers, floor-to-slab wall counteraction, above-ceiling fire barriers, etc.

5) Show the limits of construction to the contractors, and inform the contractors of existing conditions and new work in sufficient detail that they can create reasonably accurate estimates of construction costs and provide competitive bids for the work.

6) Show special construction circumstances, such as work or items to be provided and/or installed by the Owner, existing work to remain, be repaired or removed, phased construction and future additions.

REFLECTED CEILING PLANS -- CONTENT CHECKLIST

BASELINE CONTENT

Baseline content is that which is required for building department approval and construction either by an owner-builder or through a negotiated contract with an owner-selected contractor.

__ Columns, posts, walls, ceiling-high partitions

__ Suspended ceiling grids

__ Integrated ceilings

__ Exposed ceiling beams, girders, joists

__ Ceiling construction and finish

__ Furred ceilings and soffits

__ Slopes of ceiling

__ Ceiling mounted fixtures

Notes:

____

____

____

____

____

STANDARD CONTENT

Standard content is that typically required for building department approval, construction cost estimating and bidding, and complete construction by a qualified general contractor.

Includes previously listed content plus:

__ Recesses

__ Valances and detail keys

__ Overhead door and folding partition tracks and detail keys

__ Noise barriers at low partition lines

__ Ceiling-mounted casework and detail keys

__ Ceiling mounted alarms, TV cameras, other special electric fixtures

__ Access panels and detail keys

__ Chases, chutes, future shafts and detail keys

Notes:

____

____

____

____

____

____

EXTENDED CONTENT

Extended content is that required for projects requiring extra design office attention because of complex construction, elaborate detailing, and/or extended collaboration with specialized consultants.

Includes previously listed content plus:

__ Dash line indications of framing above ceiling

__ Dash line indications of hatches, wells, shafts, etc. above ceiling

__ Fire sprinkler plan

__ Ceiling heating systems

__ Exposed ductwork

Notes:

____

____

____

____

____

____

____

SCHEDULES CONTENT CHECKLIST

BASELINE CONTENT

Baseline content is that which is required for building department approval and construction either by an owner-builder or through a negotiated contract with an owner-selected contractor.

__ Notes on plan identifying finishes, door types and sizes, and window types and sizes

Notes:

____

____

STANDARD CONTENT

Standard content is that typically required for building department approval, construction cost estimating and bidding, and complete construction by a qualified general contractor.

Includes previously listed content plus:

__ Schedules as lists identifying generic substructure and finishes with references to specifications

__ Door and window symbols referenced to door and window frame schedules which are referenced in turn to detail keys showing sill, jamb, and head conditions

Notes:

____

____

EXTENDED CONTENT

Extended content is that required for projects requiring extra design office attention because of complex construction, elaborate detailing, and/or extended collaboration with specialized consultants.

Includes previously listed content plus:

__ General note explaining schedule system

__ Photo details of textures

__ Photocopy illustrations of manufactured doors and windows

__ Hardware schedule with door schedule

__ Short-hand door and window schedules included on floor plan sheets for convenient contractor reference

Large, full sheet schedules are not necessarily more comprehensive, especially if they have highly repetitive elements. Simpler schedule formats that fully convey all materials combinations in each room, keyed to a simple symbol, are preferred.

Notes:

____

____

____

FOUNDATION PLANS

THE OBJECTIVE is to create foundation plan information that will:

1) Guide our engineering consultants -- especially civil and structural engineers -- in creating actual final structural foundation drawings to guide the grading and foundation contractors.

2) If we're doing the structural framing and foundation drawings, the drawings are to guide the contractor in soil excavation, grading, preparation and form work, reinforcement placement, and concrete work.

3) Show directions and extent of framing and detail keys. Show special detail keys as required for regional requirements such as resistance to lateral forces in seismic regions or below-grade insulation in cold climates.

4) Convey information about the building sufficient to satisfy regulatory agencies such as the building department. Give particular attention to potentially hazardous soil conditions.

5) Show the limits of construction to the contractors, and inform the contractors of existing conditions and new work in sufficient detail that they can create reasonably accurate estimates of construction costs and provide competitive bids for the work.

6) Show special construction circumstances, such as work or items to be provided and/or installed by the Owner, existing work to remain, be repaired or removed, phased construction and future additions.

DETAILS CONTENT CHECKLIST

BASELINE CONTENT

Baseline content is that which is required for building department approval and construction either by an owner-builder or through a negotiated contract with an owner-selected contractor.

__ Details provided primarily only as required by building department

__ Special condition details as required for building safety, waterproofing, and solidity of connections

__ Standard details as suited to illustrate typical construction

__ Detail drawings in profile outline with materials indications and materials identification notes

Notes:

____

____

____

____

____

____

STANDARD CONTENT

Standard content is that typically required for building department approval, construction cost estimating and bidding, and complete construction by a qualified general contractor.

Includes previously listed content plus:

__ Plan and/or vertical cross sections of:
  __ Openings and penetrations
  __ Connections between different types of construction
  __ Joint connections between the same materials
  __ Profiles of ready-made products or fabrications connected to construction

__ Isometric or multi-view details where needed to clarify three dimensional detail situations

__ Details with assembly notes as needed for clarification and where not in conflict or duplication of specifications

__ General reference notes to related construction, specifications, and to drawings of other disciplines

Notes:

___

___

___

___

___

EXTENDED CONTENT

Extended content is that required for projects requiring extra design office attention because of complex construction, elaborate detailing, and/or extended collaboration with specialized consultants.

Includes previously listed content plus:

__ Specific reference notes to related drawings and specification section reference notes

__ Photodetails of existing conditions or of actual similar or related construction

__ Reference to detail construction mock-ups

Notes:

____

____

____

____

____

____

____

____

UNIFORM PRODUCTION MANAGEMENT CHECKLIST

What follows is an attempt to systematize and unify the project management process through a method we call Management by Checklist.

Admittedly, reliance on checklists over and above common sense and the need to push the work out, is not uncommon.

But a checklist can be an enormous timesaver in planning work, a great delegation and monitoring tool, an team communications medium, and an excellent instruction guide to junior staff who have no way of knowing the immense number of tasks involved in project and production management.

ADMINISTRATION -- UPDATES AFTER DESIGN DEVELOPMENT

__ Make a calendar schedule for future time, budget, and progress reviews.

__ Review previously scheduled dates for the working drawing phases. Revise the schedule as needed.

__ Update the project planning chart.

__ Bring project records up to date by recording all pertinent discussions and decisions from the previous phase that haven't yet been recorded.

__ Update contact names, phone numbers, addresses, remarks, etc. in the Project Directory.

__ Input Project Directory updates into the office-wide Project Directory data base.

Notes:

___

___

___

___

___

___

___

ADMINISTRATION -- PROJECT FINANCIAL MANAGEMENT

__ Update the project office cost records.

__ Update the project's future work-hour and cost projections.

__ Review personnel allocation in light of latest money and time budgets.

__ Confirm the client's written approval of the final design development documents and that there is written agreement to proceed with the Construction Documents.

__ Confirm the monthly statement submittal schedule and format with the client's bookkeeper.

__ Confirm the format and substantiating data required for submittal of monthly reimbursable statements.

__ Establish a schedule for documenting job costs in order to expedite submittals for payment to client.

Notes:

____

____

____

____

____

ADMINISTRATION -- SCHEDULING AND PERSONNEL ALLOCATION

__ Confirm personnel previously scheduled for the working drawing phase. Schedule hiring as needed for future phases.

__ Estimate the final number of working drawing sheets. Calculate the average total allowable work hours per sheet based on the available fee.

__ Establish a clear chain of responsibility and command for the Construction Document phase. Confirm that no employee has more than one supervisor. Distribute a memo to all parties concerned.

__ Make a master list of personnel construction document assignments.

__ Schedule training sessions for personnel who are not experi enced in special systematic production methods you use.

Notes:

____

____

____

____

____

____

UNIFORM PRODUCTION MANAGEMENT CHECKLIST
continued

ADMINISTRATION -- WORKING DRAWING PLANNING

__ Verify that final design development and presentation drawings are available for reuse in the working drawing phase.

__ Review the working drawing sheet size, sheet module, and title block design. Confirm that the title block meets all special requirements of the client and of regulatory agencies.

__ Do an index of project drawings--architectural and consultant drawings. Decide the drawing numbering system:

- __ Divisions by discipline.
- __ CSI-related divisions.
- __ Construction sequence divisions.

__ Do a one-fourth size sheet mini-mockup of all project drawings with sketches and/or notes of the data to go on each sheet. Distribute copies to concerned parties for review, then distribute final copies to all staff as a supervisory guide.

__ Decide the final printing system:

- __ Print full-size or 1/2-size diazo, or both.
- __ Print full-size or 1/2-size electrostatic copies.
- __ Create originals at small scale as "full-size mini's."
- __ Print on one or both sides of print sheets.
- __ Print offset, black and white.
- __ Print offset, multicolor.
- __ Screen background information.

__ Decide specific drafting systems appropriate to parts or all of the project, and indicate them in the mini-mockup set.

UNIFORM PRODUCTION MANAGEMENT CHECKLIST
continued

__ Micro- or minicomputer text and/or CADD graphics.
__ Functional/simplified drafting.
__ Photodrafting.
__ Machine-made or paste-up titles.
__ Typed or computer printout notation.
__ Keynotes.
__ Standard notes.
__ Standard details.
__ Linked notes and/or details with CSI numbers.
__ Full sheets of reusable standard or typical file data.
__ Paste-up.
__ Freehand drafting.
__ Tape drafting.
__ Ink drafting.
__ Enlargement/reduction copying.
__ Tape drafting.
__ Stickybacks.
__ Base sheets and overlays.
__ Screened or solid line background sheets.

__ Identify the small-scope data that will be repeated in various drawings and should be copied in multiple for paste-up. Show the small-scope repetitive elements on the miniature working drawing mockup set.

__ Identify the large-scope data on working drawings that will be repeated in various sheets; identify them as "fixed" data suited to Base Sheets. Identify "variable" data, such as the different engineers' drawings that will be combined with architectural plans, and note them as Overlay Sheets.

__ Make a separate mockup of transparency sketches of base and overlay sheet combinations. Add base and overlay combinations to the original working drawing index, to show which base sheets and which overlays are combined to make complete final prints.

UNIFORM PRODUCTION MANAGEMENT CHECKLIST
continued

__ Review the completed working drawing index and the secondary index of base and overlay sheets with all concerned parties, for feedback and revisions.

__ Complete a matrix chart that lists the final planned working drawing sheets on one side and lists all base and overlay sheets on another. Make marks in the field of the matrix showing which base sheet and overlay(s) are combined to create each final working drawing sheet. Use codes or symbols to show which base sheets will be screened or reproduced in color in the final printing.

__ Confirm the completion of limited architectural floor plan base sheet information for consultants' use. Such plan data may include:

___ Exterior walls and fenestration.
___ Interior walls and fixed partitions.
___ Door swings.
___ Equipment requiring plumbing hookups.
___ Fixtures and equipment requiring ventilation.
___ Major electrical spaces and equipment.
___ Reflected ceiling plans.
___ Mechanical and electrical chases.

Notes:

____

____

____

____

DISCIPLINES COORDINATION AND DOCUMENT CHECKING

__ Require all consultants to do their working drawing plans and elevations according to the same scale, format, and drawing positioning as the architectural drawings.

__ Confirm that CADD layering identification codes are consistent between architectural and consultants' drawings.

__ Identify any new consultants required for this phase, and negotiate contracts.

__ Before finalizing new consultant contracts, review service and contract terms with the client and obtain written client approval.

__ Transmit updated information on building occupancies to consultants; make sure the architectural design team has the identical updated information.

__ Obtain an update of the consultants' estimates of building operating costs.

__ Review with the client the consultants' building operating cost estimates; obtain from the client written approval of the proposed mechanical and electrical systems.

__ Schedule group meetings to allow consultants to compare their drawings with one another. If interferences and contradictions can't be worked out on the spot, list the problems and schedule later meetings or calls to deal with them.

UNIFORM PRODUCTION MANAGEMENT CHECKLIST
continued

__ Review previous decisions on structural, construction, mechanical, and other systems for possible economies and improvements.

__ Confirm that the various selected engineering and construction systems are compatible with one another.

__ Obtain updated estimates of spatial requirements for appurtenances and engineered systems.

__ Confirm that consultants, client, or others are handling the acquisition of approvals and permits for all utility services.

___ Gas.
___ Electric.
___ Water.
___ Sewer.
___ Telephone.
___ Cable TV.
___ Computer link.
___ Utility-supplied steam or other heating medium.
___ Utility-supplied cooling medium.

__ Obtain or update lists of special building equipment and fixtures required by the client that may affect consultants' work. Distribute the lists to the appropriate consultants.

Notes:

____

____

____

____

AGENCY CONSULTING, REVIEW, AND APPROVALS

__ Establish a checklist and timetable for the client's applications for approvals and permits.

__ Continue and update the data gathering for the checklists on regulatory agencies and codes.

OWNER-SUPPLIED DATA COORDINATION

__ Reconfirm the program's functional, occupancy, and spatial requirements with the client.

__ Compare the developed design with the client's budget. Confirm the budget agreement or settle any contradictions between stated program needs and available funding.

__ Confirm client preferences or requirements for types of construction bidding and contracting that might affect the format of construction drawings and specifications.

__ Identify possible or definite bid alternates and plan the content and organization of bid documents accordingly.

Notes:

____

____

____

____

ARCHITECTURAL DESIGN AND DOCUMENTATION

Also see ADMINISTRATION -- WORKING DRAWING PLANNING.

__ Review any changes in the program, and note their possible impact on the project design.

__ Review the Design Development documents, updates of the design, and changes in the program for possible violations of codes and regulations.

__ Review the Design Development documents, updates of the design, and changes in the program for possible conflicts with the original design intent or with fundamental engineering decisions.

__ If there are significant differences between the present design and previous design and engineering decisions, verify and document the reasons for and sources of the differences.

__ Submit a memo to all involved parties outlining the current status of work and the schedule for the Construction Documents phase.

__ Verify that all involved parties have received completely up-to-date program and schematic design data. Retrieve or otherwise remove all hold-over, obsolete design and program information.

__ If there are changes in design staff between the Design Development phase and the Construction Document phase, confirm that new staff members have acquired and assimilated previous design data and understand the reasons for the present design solution.

UNIFORM PRODUCTION MANAGEMENT CHECKLIST
continued

__ Confirm the type of construction contract to be used, such as single or separate contracts, and evaluate the effect of the contract type on drawing and specifications content and format.

__ Prepare and coordinate final specifications.

__ Review architectural working drawings as they are in process and compare them with the structural, mechanical, electrical, transportation, and other consultants' drawings by means of transparency overlays.

__ Schedule coordination check points to confirm that the architectural production team is fully informed of the most up-to-date consultants' information.

__ Schedule dates to periodically compare the work as it has developed during the working drawing phase, with budget, program, and regulatory requirements. Note any changes in building area, siting, structure, mechanical systems, construction systems, and materials that have occurred.

__ Determine and note reasons for changes in the design. Review questionable changes with those who initiated them.

__ Review preferred construction methods for impact on design and documentation.

__ Prepare data on costs and availability of special equipment and furnishings.

__ Confirm with the client whether a detailed construction cost estimate, such as a quantity survey, is desired with the final working drawings. (A detailed cost estimate, as opposed to the "Statement of Probable Construction Cost," is charged as an additional service and is highly recommended.)

__ Confirm the date for submittal of all construction documents (drawings, calculations, contracts, specifications, and up dates on construction cost estimates) to the client.

Notes:

____

____

____

____

____

____

____

____

____

____

____

____

____

UNIFORM PRODUCTION MANAGEMENT CHECKLIST
continued

STRUCTURAL DESIGN AND DOCUMENTATION

__ Schedule phases of structural engineering document production and structural/architectural coordination meetings.

__ Schedule structural, mechanical, civil, and architectural drawing cross-checking meetings.

__ Review and reach agreement with the structural engineer on the number and content of structural Construction Documents.

___ Design criteria.
___ Structural grid or system.
___ Structural framing plan(s) and sections(s).
___ Foundation plan.
___ Calculations.
___ Required clearances for other work.
___ Structural details.
___ Materials schedules.
___ Specifications.

__ Schedule completion dates for interim and final structural working drawings and specifications.

__ Confirm with the structural engineer that the proposed structural system satisfies all legal requirements.

Notes:

____

____

____

MECHANICAL DESIGN AND DOCUMENTATION

__ Establish mechanical documents production phases and dates for mechanical/architectural coordination meetings.

__ Schedule mechanical, structural, and architectural drawing cross-checking meetings.

__ Confirm with the mechanical consultant the acquisition of necessary approvals and permits for all utility services.

- ___ Gas.
- ___ Water.
- ___ Sewer.
- ___ Utility supplied steam or other heating medium.
- ___ Utility supplied cooling medium.

__ Review and reach agreement with the mechanical engineer on the number and content of final mechanical construction documents.

___ Building plans, sections, and other drawings to show:

- ___ Noise and vibration control.
- ___ HVAC system type(s) and standard(s).
- ___ Fire protection system(s).
- ___ Plumbing supply and drain types and standards.
- ___ Equipment sizes and locations.
- ___ Chase sizes and locations.
- ___ Duct sizes and locations.
- ___ Mechanical equipment spatial requirements in plan.
- ___ Mechanical equipment spatial requirements in section.
- ___ Mechanical fixture and equipment schedules.
  - ___ Mechanical construction details.

UNIFORM PRODUCTION MANAGEMENT CHECKLIST
continued

___ HVAC heat load and cooling calculations.
___ Energy use and conservation calculations.
___ Equipment and materials schedules.
___ Specifications.
___ Mechanical systems operations and maintenance instructions.

__ Confirm with the mechanical consultant the compliance of the building mechanical and plumbing system design with codes and utility company requirements.

__ Identify changes in the scope of mechanical work that have occurred during the Design Development Phase.

__ Determine the impact on cost of revisions in mechanical work.

__ Confirm with the mechanical consultant the compliance of the building mechanical and plumbing system design with codes and utility company requirements.

__ Acquire estimates for probable construction costs of the building's mechanical systems.

__ Acquire estimates for probable operating costs of the building's mechanical systems.

Notes:

____

____

____

ELECTRICAL DESIGN AND DOCUMENTATION

__ Schedule electrical documents production phases and dates for electrical/architectural coordination meetings.

__ Schedule multidiscipline and architectural drawing cross-checking meetings.

__ Identify changes in the scope of electrical work that have occurred during the Design Development Phase.

__ Determine the impact on cost of revisions in electrical work.

__ Confirm that changes in the electrical design comply with legal requirements.

__ Review and reach agreement with the electrical engineer on the number and content of Electrical Construction Documents.

- __ Building plans and sections to show:
  - __ Reflected ceiling lighting plans.
    - __ Power and switching.
    - __ Fire detection and alarm systems.
    - __ Security system.
    - __ Communications equipment, chases, and outlets.
    - __ Electrical equipment sizes, locations, and capacities.
    - __ Electrical vaults, transformer rooms.
    - __ Chase sizes and locations.
    - __ Duct sizes and locations.
    - __ Fixture schedules.
    - __ Electrical construction details.

___ Electrical, communications, security, fire, and related systems and equipment maintenance instructions.

___ Specifications.

__ Arrange the assistance of the electrical engineer in obtaining approvals and permits for electrical and communications services.

__ Obtain updated final estimates for probable electrical systems construction costs.

Notes:

____

____

____

____

____

____

____

____

____

____

CIVIL DESIGN AND DOCUMENTATION

__ Confirm that results of all previously requested site tests have been received and transmitted to the client, consultants, and the design team.

__ Identify additional tests that may be required.

__ Update the Test Log and file.

__ Schedule production phases and dates for civil/architectural coordination meetings.

__ Schedule civil, structural, landscaping, and architectural drawing cross-checking meetings.

__ Identify changes in the scope of civil engineering construction that have occurred through the Design Development Phase.

__ Determine the impact on cost of revisions in civil work.

__ Confirm that changes in the civil engineering design comply with legal requirements.

__ Review and reach agreement with the civil engineer on the number and content of civil engineering Construction Documents.

___ Site plans and sections to show:

___ Cut and fill.
___ Excavations.
___ Irrigation.
___ Drainage.
___ Site-related construction.
___ Civil engineering construction details.

UNIFORM PRODUCTION MANAGEMENT CHECKLIST
continued

___ Specifications.

__ Schedule completion dates for interim and final civil working drawings and specifications.

__ Check and confirm compliance of sitework and civil engineering design with codes and regulations.

__ Acquire updated estimates for probable civil engineering-related construction costs.

Notes:

___

___

___

___

___

___

___

___

___

___

___

LANDSCAPE DESIGN AND DOCUMENTATION

__ Schedule landscaping documents production phases and landscaping/architectural coordination meetings.

__ Schedule multidiscipline drawing cross-checking procedures or meetings.

__ Review and reach agreement with the landscape architect on the number and content of Landscape Construction Documents:

- __ Landscaping plans.
- __ Sitework construction details.
- __ Site-related plumbing work.
- __ Site-related electrical work.
- __ Specifications.
- __ Landscaping care instructions.

__ Identify special-order planting that must be ordered early, to assure delivery and installation before the completion date.

__ Schedule completion dates for interim and final landscape working drawings and specifications.

__ Update estimates for probable landscaping development costs.

Notes:

____

____

____

UNIFORM PRODUCTION MANAGEMENT CHECKLIST
continued

INTERIOR DESIGN AND DOCUMENTATION

__ Establish production phases and schedule interior design/architectural coordination meetings.

__ List and schedule special-order furnishings (such as carpet) that must be ordered early, to assure delivery and installation before the move-in date.

__ Review and reach agreement with the interior designer on the number and content of interior Construction Documents.

- __ Interior partition landscaping.
- __ Furniture selection and planning.
- __ Fixtures selection and finishes palette.
- __ Materials and finishes palette.
- __ Color schedule.
- __ Interior design detailing.
- __ Specifications.
- __ Furnishings and finish material maintenance and cleaning instructions.

__ Schedule completion dates for the final interior drawings and specifications.

__ Update estimates for probable costs of interior design furnishings and fixtures.

Notes:

____

____

____

UNIFORM PRODUCTION MANAGEMENT CHECKLIST
continued

PROJECT DEVELOPMENT SCHEDULING

__ Create or update the job calendar of estimated phase starts and completions.

__ Construction Documents.

__ 10/20% review, independent quality control check.
__ Phase 1 @ ______ % completion.
__ Phase 2 @ ______ % completion.
__ 50% review, independent midpoint quality control check.
__ Phase 3 @ ______ % completion.
__ 80/90% review, independent quality control check.

__ Final checking and completion phase.
__ Bidding/Negotiation.
__ Contract Administration.
__ Post-construction.

__ Create a schedule for job budget and progress reviews.

__ Distribute copies of the new or updated job calendar to all job participants.

Notes:

____

____

____

ESTIMATING PROBABLE CONSTRUCTION COST

__ Obtain all consultants' final construction cost estimates.

__ Prepare an overall construction estimate of probable construction costs, with a clearly stated contingency factor.

PRESENTATIONS

__ List and schedule all Construction Document presentations:

  __ Interim presentations to client.

  __ Presentations to financing agencies.

  __ Presentations to regulatory agencies.

__ Review possible future client uses of working drawing material, such as base sheet floor plans for promotional graphics.

__ Review possible office uses of working drawing material for publicity, office brochure, presentation to other client prospects, etc.

__ Prepare presentation data on preferred construction methods.

__ Identify any last minute changes in the design required by the client.

__ Note any extended repercussions from design changes, and review with the client any extensions of the Scope of Work and any required changes in design service time and cost.

UNIFORM PRODUCTION MANAGEMENT CHECKLIST
continued

__ Identify any contradictions between requested design changes and the original design program or prior client/designer decisions. Review these with the client.

__ Obtain the client's written agreement to proceed with the next phase: PRE-BIDDING, BIDDING, AND NEGOTIATIONS.

__ Prepare and submit final billing for this phase of work as per the design service contract.

Notes:

___

___

___

___

___

___

___

___

___

___

___

APPENDIX

SAMPLE STANDARD KEYNOTE SYSTEM

What follows is a prototype master set of keynotes with reference numbers and CSI specifications coordination numbers.

The reference number on the left is the location number that is placed in the field of the drawing to identify a material or assembly. The five-digit number on the right identifies the relevant specifications number and, as appropriate, the standard detail file number.

The system shown here is a sample suitable for housing construction.

Before long extensive standard keynote systems for larger construction will be available to link with your CADD system. The keynoting process will follow a uniform national standard and automated as part of the drafting process.

SITEWORK REFERENCE NOTES

1	PROPERTY LINE
2	BENCH MARK
3	SETBACK LINE
4	EASEMENT
5	LINE OF OVERHANG
6	BUILDING LINE
21	EXISTING TO BE REMOVED
22	EXISTING TO REMAIN
23	EXISTING TO BE REPAIRED
31	SLOPE TO DRAIN
32	CONSTRUCTION JOINT
33	NONSKID SURFACE

RESIDENTIAL SITEWORK KEYNOTES

REF #	KEYNOTE	CSI #
	GRADING, TRENCHING, & DRAINAGE	
2.1	EXISTING GRADE	02210
2.2	NEW FINISH GRADE	02210
2.3	COMPACTED FILL	02220
2.4	EROSION CONTROL	02270
2.5	DRAIN	02400
2.6	SUBDRAIN	02410
2.7	GUTTER DRAINAGE	02410
2.8	TRENCH DRAIN	02410
2.9	DRAINAGE FLUME/INLET	02420
2.10	CATCH BASIN	02431
2.11	STORM DRAIN	02431
2.12	SPLASH BLOCK	02435

SAMPLE STANDARD KEYNOTES (RESIDENTIAL) continued

SITE IMPROVEMENTS

2.15	FENCE	02440
2.16	SPRINKLER	02441
2.17	FOUNTAIN	02443
2.21	CHAIN LINK FENCE	02444
2.22	WIRE FENCE	02445
2.23	WOOD FENCE	02446
2.24	METAL FENCE	02447
2.25	MASONRY YARD WALL (04200)	02449
2.26	CHAIN BARRIER	02450
2.27	GUARDRAIL	02451
2.28	PARKING BUMPER	02456
2.29	RAMP	02458
2.30	PLANTER	02474
2.31	CONCRETE RET. WALL	02478
2.32	STONE RETAINING WALL	02478
2.33	CONC. BLOCK RET. WALL	02478
2.34	WOOD RETAINING WALL	02479
2.35	WOOD PLANTER	02479

LANDSCAPING

2.41	TREES TO RELOCATE	02481
2.42	SHRUBS TO RELOCATE	02481
2.43	LAWN/GRASS	02485
2.44	NEW TREES	02491
2.45	TREE PLANTERS	02491
2.46	NEW SHRUBS	02492
2.47	NEW PLANTING	02493
2.48	NEW GROUND COVER	02494
2.49	AGGREGATE PLANT. BED	02495
2.50	WOOD CHIP PLANTING BED	02496
2.51	LATH HOUSE	02498
2.52	GAZEBO	02499

PAVING & CURBS

2.55	ASPHALTIC CONCRETE	02513
2.56	ASPHALT PAVING	02513
2.57	ASPHALT SPEED BUMP	02513
2.58	BRICK PAVING	02514
2.59	BRICK WALK	02514
2.60	CONCRETE PAVING (03308)	02515
2.61	STONE PAVING	02517
2.62	CONCRETE PAVERS	02518
2.63	GRAVEL	02519
2.64	WOOD CURB	02521

SAMPLE STANDARD KEYNOTES (RESIDENTIAL) continued

SITEWORK KEYNOTES continued

PAVING & CURBS continued

2.65	WOOD EDGE STRIP	02522
2.66	WOOD STEPS	02523
2.67	WOOD TIMBER PLANTER	02523
2.68	STONE CURBS	02524
2.69	PRECAST CONCRETE CURB	02526
2.70	CONCRETE CURB	02528
2.71	GUTTER	02528
2.72	CONCRETE WALK/PAVING	02529
2.73	CONCRETE STEPS	02529

UTILITIES

2.81	HOSE BIBB	02640
2.82	CISTERN	02665
2.83	WATER SUPPLY MAIN	02665
2.84	WATER METER (15181)	02665
2.85	WATER WELL	02670
2.86	GAS MAIN	02685
2.87	GAS METER (15182)	02685
2.88	GAS SHUTOFF VALVE	02685
2.89	SEWER MAIN	02730
2.90	SEWER LINE FROM HOUSE	02730
2.91	SEPTIC SYSTEM	02740

DIVISION 3--CONCRETE

3.1	CONCRETE SLAB	03300
3.2	CONCRETE WALL	03351

DIVISION 4--MASONRY

4.1	BRICK PLANTER	04203
4.2	BRICK WALL	04210
4.3	ADOBE WALL	04212
4.4	CONC. BLOCK RET. WALL	04228
4.5	CONC. BLOCK WALL	04229
4.6	RUBBLE STONE	04400
4.7	STONE WALL	04410

SAMPLE STANDARD KEYNOTES (RESIDENTIAL) continued

SITEWORK KEYNOTES continued

DIVISION 5--METALS

5.1	METAL POST	05120
5.2	PIPE COLUMN	05120
5.3	METAL STAIR	05511
5.4	METAL HANDRAIL	05521
5.5	METAL BALCONY	05522
5.6	METAL GRATING	05530

DIVISION 6--WOOD

6.1	WOOD POST	06100
6.2	WOOD DECK	06125
6.3	WOOD BALCONY	06126
6.4	WOOD STAIR	06430
6.5	WOOD HANDRAIL	06430
6.6	WOOD TRELLIS	06450

DIVISION 9--FINISHES

9.1	TILE PAVING	09300
9.2	TILE FINISH STEPS	09300

DIVISION 10--SPECIALTIES

10.1	COVERED WALKWAY	10531
10.2	CAR SHELTER	10532
10.3	MAILBOX	10552

DIVISION 13--SPECIAL CONSTRUCTION

13.1	GREENHOUSE	13123
13.2	GRANDSTANDS/BLEACHERS	13125
13.3	SWIMMING POOL	13151
13.4	WADING POOL	13151
13.5	JACUZZI	13153
13.6	HOT TUB	13154
13.7	LIQUID/GAS STORAGE TANK	13410
13.8	SOLAR ENERGY SYSTEM	13600

SAMPLE STANDARD KEYNOTES (RESIDENTIAL) continued

SITEWORK KEYNOTES continued

DIVISION 15--MECHANICAL

15.1	WATER SHUTOFF BOX	15181
15.2	FIRE HYDRANT	15530

DIVISION 16--ELECTRICAL

16.1	POWER POLE	16050
16.2	ELECTRICAL BURIED CABLE	16400
16.3	ELECT. OVERHEAD CABLE	16400
16.4	ELECT.SERVICE ENTRANCE	16420
16.5	ELECTRIC METER	16430
16.6	EXTERIOR LIGHTING	16520
16.7	EXTERIOR ELECT.L OUTLET	16521
16.8	LIGHT STANDARDS	16530
16.9	WALKWAY LIGHTS	16530
16.10	TELEPHONE POLE	16700
16.11	TELEPHONE BURIED CABLE	16740
16.12	TELE. OVERHEAD CABLE	16740
16.13	TV OVERHEAD CABLE	16780
16.14	TV BURIED CABLE	16780
16.15	DISH ANTENNA	16785

SAMPLE STANDARD KEYNOTES (RESIDENTIAL) continued

FLOOR PLAN KEYNOTES

REF #	KEYNOTE	CSI #
	DIVISION 3--CONCRETE	
3.1	CONCRETE SLAB	03300
3.3	CONCRETE WALL	03302
	DIVISION 4--MASONRY	
4.1	BRICK	04210
4.3	FACE BRICK	04215
4.5	CONCRETE BLOCK	04220
4.7	GLASS BLOCK	04270
4.9	STONE	04410
4.11	STONE VENEER	04450
4.13	MASONRY FIREPLACE	04550
4.15	HEARTH	04550
	DIVISION 5--METALS	
5.1	STEEL HEADER ABOVE	05120
5.3	STEEL BEAM ABOVE	05121
5.5	METAL POST	05123
5.7	METAL STAIR	05510
5.9	METAL RAILING	05521
5.11	HANDRAIL	05521
	DIVISION 6--WOOD & PLASTIC ROUGH CARPENTRY	
6.1	WOOD POST	06101
6.3	WOOD BEAM ABOVE	06104
6.5	PLYWOOD DIAPHRAGM SHEAR WALL	06114
6.7	WOOD HEADER	06124
6.9	WOOD DECKING	06125

SAMPLE STANDARD KEYNOTES (RESIDENTIAL) continued

FLOOR PLAN KEYNOTES continued

FINISH CARPENTRY

6.11	PEGBOARD	06255
6.13	WOOD TRIM	06220
6.15	DECORATIVE MILLWORK	06220
6.17	LAMINATED PLASTIC COUNTERTOP	06240
6.19	WOOD PANELING	06250

ARCHITECTURAL WOODWORK

6.21	WOOD CABINETWORK	06410
6.23	DISPLAY CABINET	06411
6.25	WOOD SHELVING	06412
6.27	WARDROBE POLE & SHELF	06413
6.29	CEDAR LINED CLOSET	06414
6.31	LINEN CLOSET SHELVES	06415
6.33	WOOD COUNTERTOP	06417
6.35	WOOD BASE CABINET	06418
6.37	WOOD STAIR	06431
6.39	WOOD RAILING	06440
6.41	WOOD TRELLIS/LATTICE	06450
6.43	WORKBENCH	06453
6.45	WOOD SCREEN	06570

DIVISION 7--THERMAL & MOISTURE PROTECTION

7.1	WATERPROOF MEMBRANE	07110
7.3	WATERPROOF UNDERLAYMENT AT BATHROOM FLOORS	07113
7.5	INSULATION	07210
7.7	SKYLIGHT ABOVE	07800
7.9	EXPANSION JOINT/SEALANT	07900

DIVISION 8--DOORS & WINDOWS

8.1	ACCESS PANEL DOOR	08305
8.3	GARAGE DOOR	08360
8.5	SLIDING DOOR TRACK	08710
8.7	SADDLE	08711
8.9	WEATHERSTRIPPING	08712
8.11	SLIDING DOOR POCKET	08730
8.13	GLAZING	08800
8.15	FIXED GLASS	08810
8.17	OBSCURE GLASS	08811
8.19	VIEW PANEL	08821

SAMPLE STANDARD KEYNOTES (RESIDENTIAL) continued

FLOOR PLAN KEYNOTES continued

DIVISION 9--FINISHES

SEE FINISH SCHEDULE FOR COMPLETE LISTS AND DESCRIPTIONS OF FINISHED FLOOR, WALL AND CEILING CONSTRUCTION.

9.1	PLASTER FURRED WALL	09200
9.3	SUSPENDED CLG. ABOVE	09120
9.5	GYPSUM LATH & PLASTER	09201
9.7	METAL LATH & PLASTER	09203
9.9	GYPSUM WALLBOARD	09250
9.11	WALLBOARD FURRING	09263
9.13	TILE	09300
9.15	QUARRY TILE	09330
9.17	STONE TILE	09380
9.19	TERRAZZO	09410
9.21	ACOUSTICAL INSULATION	09530
9.23	SOUND ISOLATION WALL	09530

DIVISION 10--SPECIALTIES

10.1	GRILLE/SCREEN	10230
10.3	SERVICE WALL ACCESS	10250
10.5	PREFAB FIREPLACE	10301
10.7	MAIL SLOT	10551
10.9	MAIL BOX	10552
10.11	FULL LENGTH MIRROR	10800
10.13	MAKE-UP MIRROR	10800
10.15	MEDICINE CABINET	10820
10.17	TUB/SHOWER ENCLOSURE	10825
10.19	GRAB BAR	10830
10.21	PREFAB WARDROBE	10900
10.23	PREFAB SHELF & POLE	10910
10.25	CLOTHES HOOK	10920

DIVISION 11--EQUIPMENT

11.1	REFRIGERATOR	11405
11.3	FREEZER	11407
11.5	WET BAR	11410
11.7	DINING BAR	11415
11.9	SINK	11417
11.11	STOVE	11420
11.13	OVEN	11423
11.15	COOKTOP	11425
11.17	MICROWAVE	11427
11.19	DISHWASHER	11440

SAMPLE STANDARD KEYNOTES (RESIDENTIAL) continued

FLOOR PLAN KEYNOTES continued

DIVISION 11--EQUIPMENT continued

11.21	COMPACTOR	11441
11.23	CLOTHES WASHER	11450
11.25	CLOTHES DRYER	11450
11.27	DRYER VENT TO OUTSIDE	11451
11.29	BUILT-IN IRONING BOARD	11454

DIVISION 12--FURNISHINGS

12.1	PREFAB CASEWORK	12370
12.3	DRAPERY POCKET	12501
12.5	DRAPE/BLIND VALANCE	12503
12.7	BLINDS	12510
12.9	INTERIOR SHUTTERS	12527
12.11	BUILT IN SEATING	12623
12.13	INTERIOR PLANTER	12820

DIVISION 13--SPECIAL CONSTRUCTION

13.1	JACUZZI	13153
13.3	HOT TUB	13154

DIVISION 14--CONVEYING SYSTEMS

14.1	DUMBWAITER	14100
14.3	LAUNDRY CHUTE	14560

DIVISION 15--MECHANICAL

15.1	PLUMBING CHASE	15400
15.3	FLOOR DRAIN	15421
15.5	WATER HEATER	15450
15.7	AIR CONDITIONER	15500
15.9	FURNACE	15610
15.11	FAN TO OUTSIDE	15941

DIVISION 16--ELECTRICAL

16.1	ELECTRIC PANELBOARD	16160

SAMPLE STANDARD KEYNOTES (RESIDENTIAL) continued

EXTEROR ELEVATION REFERENCE NOTES

1 PROPERTY LINE
2 BENCH MARK
3 SETBACK LINE
4 EASEMENT
5 LINE OF OVERHANG
6 BUILDING LINE

21 EXISTING TO BE REMOVED
22 EXISTING TO REMAIN
23 EXISTING TO BE REPAIRED

15 SLOPE TO DRAIN
16 CONSTRUCTION JOINT

EXTEROR ELEVATION KEYNOTES

REF #	KEYNOTE	CSI #
	DIVISION 2--SITEWORK	
2.1	EXISTING GRADE	02210
2.2	NEW FINISH GRADE	02210
2.3	COMPACTED FILL	02220
2.4	PILES	02360
2.5	FOOTING DRAIN	02411
2.6	UNDERSLAB DRAIN	02412
2.7	AREA DRAIN	02420
2.8	CATCH BASIN	02431
2.9	SPLASH BLOCK	02435
2.15	FENCE	02440
2.16	CHAIN LINK FENCE	02444
2.17	WIRE FENCE	02445
2.18	WOOD FENCE	02446
2.19	METAL FENCE	02447
2.20	MASONRY YARD WALL	02449
2.21	GUARDRAIL	02451
2.22	PARKING BUMPER	02456
2.23	RAMP	02458
2.24	PLANTER	02474
2.25	RETAINING WALL	02478
2.26	CONCRETE RETAINING WALL	02478
2.27	STONE RETAINING WALL	02478
2.28	CONC. BLK RETAINING WALL	02478
2.29	WOOD RETAINING WALL	02479
2.30	WOOD PLANTER	02479
2.31	CONCRETE STEPS	02529
2.32	CONCRETE WALK/PAVING	02529

SAMPLE STANDARD KEYNOTES (RESIDENTIAL) continued

EXTERIOR ELEVATION KEYNOTES continued

DIVISION 3--CONCRETE

3.1	CONCRETE PIER	03303
3.2	CONCRETE GRADE BEAM	03304
3.3	CONCRETE FOOTING	03305
3.4	CRWL SPACE ACCESS PANEL	03305
3.5	CONCRETE FNDATION WALL	03306
3.6	CONCRETE BASEMENT WALL	03307
3.7	CONCRETE SLAB	03308
3.8	SLAB FOOTING	03308
3.9	CONCRETE COLUMN	03317
3.10	CONCRETE WALL	03351
3.11	EXPOSED AGGREGATE CONC.	03351
3.12	TOOLED CONCRETE	03352
3.13	SAND BLASTED CONCRETE	03352
3.14	PRECAST CONCRETE	03400

DIVISION 4--MASONRY

4.1	CONTROL JOINT	04180
4.2	BRICK PLANTER	04203
4.3	BRICK COLUMN	04210
4.4	BRICK MASONRY CHIMNEY	04211
4.5	ADOBE WALL	04212
4.6	BRICK VENEER	04215
4.7	WEEP HOLE	04215
4.8	BRICK WALL	04216
4.9	CONCRETE BLOCK WALL	04220
4.10	CONCRETE BLOCK PLANTER	04220
4.11	CONC. BLOCK RET. WALL	04228
4.12	CONC. BLOCK FOUND. WALL	04229
4.13	CONC. BLOCK COLUMN	04230
4.14	CERAMIC TILE VENEER	04250
4.15	GLASS BLOCK	04270
4.16	STONE WALL	04400
4.17	FIELD STONE	04410
4.18	CUT STONE	04420
4.19	MASONRY COPING	04420
4.20	STONE VENEER	04450

DIVISION 5--METALS

5.1	METAL FRAME PARAPET	05100
5.2	STEEL LINTEL	05117
5.3	STRUCTURAL STEEL	05120
5.4	STEEL TUBE POST	05122
5.5	STEEL JOISTS	05210
5.6	METAL DECKING	05310
5.8	METAL STAIR	05511

SAMPLE STANDARD KEYNOTES (RESIDENTIAL) continued

EXTERIOR ELEVATION KEYNOTES continued

DIVISION 5--METALS continued

5.8	METAL RAILING	05521
5.9	METAL BALCONY	05522
5.10	WEATHER VANE	05700
5.11	ORNAMENTAL STAIR	05710
5.12	ORNAMENTAL RAILING	05720
5.13	METAL TRELLIS	05725
5.14	METAL SOFFIT	05730
5.15	METAL FASCIA	05730

DIVISION 6--WOOD ROUGH CARPENTRY

6.1	WOOD COLUMN	06101
6.2	WOOD POST	06101
6.3	WOOD GIRDER	06103
6.4	WOOD BEAM	06104
6.5	WOOD JOIST	06105
6.6	WOOD SHEATHING	06113
6.7	WOOD HEADER	06124
6.8	WOOD DECKING	06125
6.9	WOOD BALCONY	06126
6.10	WOOD STEPS	06127
6.11	WOOD POLE	06133
6.12	GLU-LAM GIRDER	06181
6.13	GLU-LAM BEAM	06181
6.14	WOOD TRUSS	06190

FINISH CARPENTRY

6.20	DECORATIVE MILLWORK	06220
6.21	WOOD SHUTTERS	06235
6.22	WOOD PANELING	06420
6.23	WOOD SOFFIT	06420
6.24	WOOD FASCIA	06420
6.25	WOOD STAIR	06431
6.26	WOOD RAILING	06440
6.27	WOOD GUARDRAIL	06440
6.28	WOOD TRELLIS	06450
6.29	WOOD TRIM	06450
6.30	WOOD SCREEN	06570

SAMPLE STANDARD KEYNOTES (RESIDENTIAL) continued

EXTERIOR ELEVATION KEYNOTES continued

DIVISION 7 -- THERMAL & MOISTURE PROTECTION

7.1	BELOW GRADE WATERPROOFING	07100
7.2	WATERPROOFING	07100
7.3	INSULATION	07240
7.4	ASPHALT SHINGLES	07311
7.5	WD SHINGLES (FIRE HZRD)	07313
7.6	WOOD SHAKES (FIRE HZARD)	07313
7.7	SLATE SHINGLES	07314
7.9	METAL SHINGLES	07316
7.10	CLAY TILES	07321
7.11	CEMENT TILES	07322
7.15	METAL SIDING	07411
7.16	PREFORMED METAL ROOF	07412
7.17	WOOD SIDING	07461
7.18	PLYWOOD SIDING	07465
7.19	BUILT-UP ROOFING	07510
7.20	SHEET METAL ROOFING	07610
7.21	METAL FLASHING	07620
7.22	GUTTER	07631
7.23	DOWNSPOUT	07632
7.24	SCUPPER	07633
7.25	EAVE SNOW GUARD	07635
7.26	FLEXIBLE FLASHING	07650
7.27	GRAVEL STOP	07660
7.28	METAL COPING	07665
7.29	SKYLIGHT	07810
7.30	ROOF DRAIN	07822
7.31	VENT	07825
7.32	JOINT SEALANT	07910
7.33	CAULKING	07921

DIVISION 8--DOORS & WINDOWS

SEE DOOR AND WINDOW SCHEDULE FOR SPECIFIC DOOR AND WINDOW TYPES AND FINISHES

8.1	OVERHEAD DOOR	08360
8.2	VERTICAL LIFT WOOD DOOR	08366
8.3	VERTICAL LIFT METAL DOOR	08366
8.4	SLIDING GLASS DOOR	08370
8.5	MONITOR/ROOF WINDOW	08655

SAMPLE STANDARD KEYNOTES (RESIDENTIAL) continued

EXTERIOR ELEVATION KEYNOTES continued

GLAZING

8.10	FIXED GLASS	08800
8.11	DOUBLE GLAZING	08802
8.12	TEMPERED GLASS	08813
8.13	WIRE GLASS	08814
8.14	OBSCURE GLASS	08815
8.15	SAFETY GLASS	08822
8.16	INSULATING GLASS	08823

DIVISION 9--FINISHES

SEE DIVISIONS 3, 4, 5, 6, 7 ETC. FOR EXTERIOR CONSTRUCTION AND FINISH MATERIALS

9.1	ADOBE FINISH	09225
9.2	STUCCO	09230
9.3	CERAMIC TILE	09310
9.4	QUARRY TILE	09330
9.5	STONE TILE	09332

DIVISION 10--SPECIALTIES

10.1	CRAWL SPACE VENT	10200
10.2	LOUVER	10210
10.3	VENT	10210
10.4	EAVE SOFFIT VENT	10210
10.5	SECURITY WINDOW GUARD	10240
10.6	GRILLE	10240
10.7	PREFAB CHIMNEY	10300
10.8	ADDRESS PLAQUE	10420
10.9	AWNING	10535
10.10	MAIL BOX	10552
10.11	SUN SCREEN	10750

DIVISION 13--SPECIAL CONSTRUCTION

13.1	POOL	13150
13.2	SOLAR COLLECTOR	13610
13.3	PACKAGED SOLAR SYSTEM	13640
13.4	PHOTOVLTAIC COLLECTORS	13650

SAMPLE STANDARD KEYNOTES (RESIDENTIAL) continued

EXTERIOR ELEVATION KEYNOTES continued

DIVISION 15--MECHANICAL

15.1	HOSE BIBB	15109
15.2	WALL HYDRANT	15109
15.3	GAS METER	15182
15.4	EXTERIOR FIRE SPRINKLER	15330
15.5	DOMESTIC SOLAR WATER HEATER	15431

DIVISION 16--ELECTRICAL

16.1	EXTERIOR LIGHTING	16520
16.2	LIGHT STANDARDS	16530
16.3	WALKWAY LIGHTS	16530
16.4	LIGHTNING ROD	16670
16.5	FIRE ALARM	16721
16.6	BURGLAR ALARM	16727
16.7	SURVEILLANCE TV CAMERA	16780
16.8	SATELLITE DISH ANTENNA	16781
16.9	OVERHEAD CABLE ENTRY	16401
16.10	ELECTRICAL SERVICE ENTRY	16420
16.11	LIGHTNING ROD	16601
16.12	TV CABLE ENTRY	16780
16.13	DISH ANTENNA	16781
16.14	SNOW MELTING CABLES	16858

SAMPLE STANDARD KEYNOTES (RESIDENTIAL) continued

CROSS SECTION & WALL SECTION REFERENCE NOTES

1	PROPERTY LINE
2	BENCH MARK
3	SETBACK LINE
4	EASEMENT
5	LINE OF OVERHANG
6	BUILDING LINE
21	EXISTING TO BE REMOVED
22	EXISTING TO REMAIN
23	EXISTING TO BE REPAIRED
15	SLOPE TO DRAIN
16	CONSTRUCTION JOINT

CROSS SECTION & WALL SECTION KEYNOTES

REF #	KEYNOTE	CSI #
	DIVISION 2--SITEWORK	
2.1	EXISTING GRADE	02210
2.2	NEW FINISH GRADE	02210
2.3	COMPACTED FILL	02220
2.4	PILES	02360
2.5	FOOTING DRAIN	02411
2.6	UNDERSLAB DRAIN	02412
2.15	FENCE	02440
2.20	MASONRY YARD WALL	02449
2.24	PLANTER	02474
2.25	RETAINING WALL	02478
2.26	CONCRETE RETAINING WALL	02478
2.27	STONE RETAINING WALL	02478
2.28	CONC. BLK RETAINING WALL	02478
2.29	WOOD RETAINING WALL	02479
2.30	WOOD PLANTER	02479
2.31	CONCRETE STEPS	02529
2.32	CONCRETE WALK/PAVING	02529

SAMPLE STANDARD KEYNOTES (RESIDENTIAL) continued

CROSS SECTION AND WALL SECTION KEYNOTES continued

DIVISION 3--CONCRETE

3.1	REINFORCING BARS	03210
3.2	WIRE MESH REINFORCING	03220
3.3	CONC. SLAB CONTROL JOINT	03251
3.4	CONSTRUCTION JOINT	03252
3.5	DOWEL CONNECTION BETWEEN SLABS	03253
3.9	CONCRETE PIER	03303
3.10	CONCRETE GRADE BEAM	03304
3.11	CONCRETE FOOTING	03305
3.12	CRWL SPACE ACCESS PANEL	03305
3.13	CONCRETE FNDATION WALL	03306
3.14	CONCRETE BASEMENT WALL	03307
3.15	CONCRETE SLAB	03308
3.16	SLAB FOOTING	03308
3.17	CONCRETE COLUMN	03317
3.18	CONCRETE WALL	03351
3.19	PRECAST CONCRETE	03400
3.20	FIREPLACE FOUNDATION	03308
3.21	FURNACE SLAB	03308
3.22	SLAB FOOTING	03308
3.23	CONCRETE CURB AT SLAB	03308
3.24	GARAGE PERIMETER SLAB CURB	03308
3.25	GARAGE SLAB SLOPED TO APRON	03308
3.26	GIRDER POCKET	03310
3.27	FOUNDATION WALL LIP FOR MASONRY VENEER	03310

SAMPLE STANDARD KEYNOTES (RESIDENTIAL) continued

CROSS SECTION AND WALL SECTION KEYNOTES continued

DIVISION 4--MASONRY

4.1	CONTROL JOINT	04180
4.2	MASONRY PARAPET	04200
4.3	CHIMNEY COPING CAP	04200
4.4	BRICK CHIMNEY	04211
4.5	FLUE	04551
4.6	BRICK PLANTER	04203
4.7	BRICK COLUMN	04210
4.8	BRICK MASONRY CHIMNEY	04211
4.9	ADOBE WALL	04212
4.10	BRICK VENEER	04215
4.11	WEEP HOLE	04215
4.12	BRICK WALL	04216
4.13	CONCRETE BLOCK WALL	04220
4.20	CONCRETE BLOCK PLANTER	04220
4.21	CONC. BLOCK RET. WALL	04228
4.22	CONC. BLOCK FOUND. WALL	04229
4.23	CONC. BLOCK COLUMN	04230
4.24	CERAMIC TILE VENEER	04250
4.25	GLASS BLOCK	04270
4.26	STONE WALL	04400
4.27	FIELD STONE	04410
4.28	CUT STONE	04420
4.29	MASONRY COPING	04420
4.30	STONE VENEER	04450

DIVISION 5--METALS

5.1	METAL FRAME PARAPET	05100
5.2	STEEL LINTEL	05117
5.3	STRUCTURAL STEEL	05120
5.4	STEEL TUBE POST	05122
5.6	STEEL JOISTS	05210
5.7	COMPOSITE JOIST SYSTEM	05260
5.8	METAL DECKING	05310
5.9	METAL STUDS	05410
5.15	METAL STAIR	05511
5.16	METAL RAILING	05521
5.17	METAL BALCONY	05522
5.18	METAL GRATING	05530
5.21	ORNAMENTAL STAIR	05710
5.22	ORNAMENTAL RAILING	05720
5.23	METAL TRELLIS	05725
5.24	METAL SOFFIT	05730
5.25	METAL FASCIA	05730
5.30	EXPANSION JOINT	05802
5.31	EXPANSION JOINT COVER	05802

SAMPLE STANDARD KEYNOTES (RESIDENTIAL) continued

CROSS SECTION AND WALL SECTION KEYNOTES continued

DIVISION 6--WOOD ROUGH CARPENTRY

6.1	WOOD COLUMN	06101
6.2	WOOD POST	06101
6.3	WOOD GIRDER	06103
6.4	WOOD BEAM	06104
6.5	WOOD BOX BEAM	06104
6.6	WOOD CAP/POST AT CONC. PIER	06105
6.7	MUD SILL	06105
	BOTTOM PLATE	
6.8	DOUBLE TOP PLATE	
6.9	WOOD JOIST	06110
6.10	WOOD JOIST	06105
6.11	WOOD STUDS	06110
6.12	WOOD FRAME PARAPET	06110
6.13	JOIST BLOCKING	06110
6.14	DOUBLE JOISTS UNDER PARALLEL PARTITIONS	06110
6.15	DOUBLE HEADER JOISTS AT THRU-FLOOR OPENINGS	06110
6.16	TIMBER CONNECTOR	06111
6.17	FRAMING CLIP	06112
6.18	BEAM HANGER	06113
6.19	JOIST HANGER	06113
6.20	SHEAR WALL	06114
6.21	SHEATHING	06115
6.22	WOOD SHEATHING	06113
6.23	HEADER	06124
6.24	LEDGER	06124
6.25	WOOD DECKING	06125
6.26	WOOD BALCONY	06126
6.27	WOOD STEPS	06127
6.28	WOOD POLE	06133
6.29	WOOD-METAL JOISTS	06150
6.30	GLU-LAM GIRDER	06181
6.31	GLU-LAM BEAM	06181
6.32	WOOD TRUSS	06190

SAMPLE STANDARD KEYNOTES (RESIDENTIAL) continued

CROSS SECTION AND WALL SECTION KEYNOTES continued

FINISH CARPENTRY

6.20	DECORATIVE MILLWORK	06220
6.22	WOOD PANELING	06420
6.23	WOOD SOFFIT	06420
6.24	WOOD FASCIA	06420
6.25	WOOD STAIR	06431
6.26	WOOD RAILING	06440
6.27	WOOD GUARDRAIL	06440
6.28	WOOD TRELLIS	06450
6.29	WOOD TRIM	06450
6.30	WOOD SCREEN	06570

DIVISION 7--THERMAL & MOISTURE PROTECTION

7.1	BELOW GRADE WATERPROOFING	07100
7.2	WATERPROOFING	07100
7.3	MEMBRANE WTERPROOFING	07110
7.4	DAMPPROOFING	07150
7.5	VAPOR BARRIER	07190
7.6	VAPOR BARRIER WITH SAND COVER OVER CRAWL SPACE	07190
7.7	RIGID INSULATION	07212
7.8	PERIMETER THERMAL INSULATION	07212
7.9	BATT THERMAL INSULATION	07213
7.10	FOAMED-IN-PLACE INSULATION	07214
7.11	SPRAYED INSULATION	07216
7.12	GRANULAR INSULATION	07216
7.13	INSULATION	07240
7.14	FIREPROOFING	07250
7.15	FIRESTOPPING	07270
7.20	ASPHALT SHINGLES	07311
7.21	WD SHINGLES (FIRE HZARD)	07313
7.22	WOOD SHAKES (FIRE HZARD)	07313
7.23	SLATE SHINGLES	07314
7.24	METAL SHINGLES	07316
7.25	CLAY TILES	07321
7.26	CEMENT TILES	07322

SAMPLE STANDARD KEYNOTES (RESIDENTIAL) continued

CROSS SECTION AND WALL SECTION KEYNOTES continued

DIVISION 7--THERMAL & MOISTURE PROTECTION cont.

7.15	METAL SIDING	07411
7.16	PREFORMED METAL ROOF	07412
7.17	WOOD SIDING	07461
7.18	PLYWOOD SIDING	07465
7.19	BUILT-UP ROOFING	07510
7.20	SHEET METAL ROOFING	07610
7.21	METAL FLASHING	07620
7.22	GUTTER	07631
7.23	DOWNSPOUT	07632
7.24	SCUPPER	07633
7.25	EAVE SNOW GUARD	07635
7.26	FLEXIBLE FLASHING	07650
7.27	GRAVEL STOP	07660
7.28	METAL COPING	07665
7.29	SKYLIGHT	07810
7.59	SKYLIGHT CURB	07810
7.30	ROOF DRAIN	07822
7.31	VENT	07825
7.60	ROOF HATCH	07830
7.61	CABLE SUPPORT & GUY	07871
7.32	JOINT SEALANT	07910
7.33	CAULKING	07921

DIVISION 8--DOORS & WINDOWS

SEE DOOR AND WINDOW SCHEDULE FOR SPECIFIC DOOR AND WINDOW TYPES AND FINISHES

8.1	OVERHEAD DOOR	08360
8.2	VERTICAL LIFT WOOD DOOR	08366
8.3	VERTICAL LIFT METAL DOOR	08366
8.4	SLIDING GLASS DOOR	08370
8.5	MONITOR/ROOF WINDOW	08655

GLAZING

8.10	FIXED GLASS	08800
8.11	DOUBLE GLAZING	08802
8.12	TEMPERED GLASS	08813
8.13	WIRE GLASS	08814
8.14	OBSCURE GLASS	08815
8.15	SAFETY GLASS	08822
8.16	INSULATING GLASS	08823

SAMPLE STANDARD KEYNOTES (RESIDENTIAL) continued

CROSS SECTION AND WALL SECTION KEYNOTES continued

DIVISION 9--FINISHES

SEE DIVISIONS 3, 4, 5, 6, 7 ETC. FOR EXTERIOR CONSTRUCTION AND FINISH MATERIALS

9.1	ADOBE FINISH	09225
9.2	STUCCO	09230
9.3	CERAMIC TILE	09310
9.4	QUARRY TILE	09330
9.5	STONE TILE	09332

DIVISION 10--SPECIALTIES

10.1	CRAWL SPACE VENT	10200
10.2	LOUVER	10210
10.3	VENT	10210
10.4	EAVE SOFFIT VENT	10210
10.5	SECURITY WINDOW GUARD	10240
10.6	GRILLE	10240
10.7	THRU-WALL ACCESS PANEL	10250
10.8	SCUTTLE	10250
10.9	RODENTPROOFING	10292
10.10	TERMITE SHIELD	10294
10.11	PREFAB CHIMNEY	10300
10.12	FIREPLACE ASH PIT AND CLEANOUT	10300
10.13	AWNING	10535
10.14	SUN SCREEN	10750

DIVISION 13--SPECIAL CONSTRUCTION

13.1	POOL	13150
13.2	SOLAR COLLECTOR	13610
13.3	PACKAGED SOLAR SYSTEM	13640
13.4	PHOTOVLTAIC COLLECTORS	13650

SAMPLE STANDARD KEYNOTES (RESIDENTIAL) continued

CROSS SECTION AND WALL SECTION KEYNOTES continued

DIVISION 15--MECHANICAL

15.1	EXTERIOR FIRE SPRINKLER	15330
15.3	ROOF DRAIN	15422
15.4	PLUMBING CHASE	15400
15.5	SOIL VENT STACK	15405
15.6	THRU-WALL SLEEVE FOR SUPPLY PIPE	15410
15.7	THRU-WALL SLEEVE FOR WASTE DRAIN PIPE	15410
15.8	FLOOR DRAIN	15421
15.9	PLUMBING ACCESS PANEL	15423
15.12	SOLAR WATER HEATER	15431
15.13	WATER HEATER	15450
15.14	FURNACE	15610
15.15	HEAT PUMP	15770
15.20	AIR HANDLING VENT	15800
15.21	HOSE BIBB	15109
15.22	WALL HYDRANT	15109
15.23	ROOF EXHAUST FAN	15829
15.24	AIR INTAKE	15878
15.25	DUCTWORK	15890
15.26	UNDER-FLOOR PLENUM	15890
15.27	ROOF DRAIN	15422

DIVISION 16--ELECTRICAL

16.1	ELECTRICAL CHASE	16050
16.2	RACEWAYS	16110
16.3	SLEEVE FOR CONDUIT	16110
16.4	EXTERIOR LIGHTING	16520
16.5	LIGHTNING ROD	16670
16.6	FIRE ALARM	16721
16.7	BURGLAR ALARM	16727
16.8	SURVEILLANCE TV CAMERA	16780
16.9	SATELLITE DISH ANTENNA	16781
16.10	OVERHEAD CABLE ENTRY	16401
16.11	ELECTRICAL SERVICE ENTRY	16420
16.12	LIGHTNING ROD	16601
16.13	TV CABLE ENTRY	16780
16.14	DISH ANTENNA	16781
16.15	SNOW MELTING CABLES	16858

SAMPLE STANDARD KEYNOTES (RESIDENTIAL) continued

ROOF KEYNOTES

REF #	KEYNOTE	CSI #
	DIVISION 2--SITEWORK	
2-1	PROPERTY LINE	
2-3	BENCH MARK	
2-5	SETBACK LINE	
2-7	EASEMENT	
2-9	EXISTING TO BE REMOVED	
2-11	EXISTING TO REMAIN	
2-13	EXISTING TO BE REPAIRED	
2-21	STEPS	
2-22	EXTERIOR STAIRS	
2-23	RAMP	
2-25	DRIVEWAY	
2-27	WALK WAY	
2-31	PAVING	
2-33	TERRACE	
2-35	PATIO	
2-37	BALCONY	
2-39	DECKING	
2-41	NONSKID SURFACE	
2-43	SLOPE TO DRAIN	
2-45	CONSTRUCTION JOINT	
2-51	PLANTING	
2-61	LINE OF OVERHANG	
2-63	BUILDING LINE	

SAMPLE STANDARD KEYNOTES (RESIDENTIAL) continued

ROOF KEYNOTES continued

DIVISION 3--CONCRETE

3.1	CONCRETE ROOF DECK	03300
3.2	STRUCTURAL CONCRETE	03310

DIVISION 4--MASONRY

4.1	MASONRY PARAPET	04200
4.2	CHIMNEY COPING CAP	04200
4.3	BRICK CHIMNEY	04210
4.4	FLUE	04551

DIVISION 5--METALS

5.1	STEEL JOISTS	05210
5.2	COMPOSITE JOIST SYSTEM	05260
5.3	METAL DECKING	05300
5.4	ROOF LADDER	05517
5.5	METAL RAILING/HANDRAIL	05521
5.6	WEATHER VANE	05700
5.7	METAL TRELLIS	05725
5.8	METAL SOFFIT	05730
5.9	METAL FASCIA	05730
5.10	EXPANSION JOINT	05802
5.11	EXPANSION JOINT COVER	05802

DIVISION 6--WOOD & PLASTIC

ROUGH CARPENTRY

6.1	WOOD COLUMN	06101
6.2	WOOD POST	06101
6.3	WOOD GIRDER	06103
6.4	WOOD BEAM	06104
6.5	WOOD BOX BEAM	06104
6.6	WOOD JOISTS	06105
6.7	WOOD RAFTERS	06106
6.8	WOOD FRAME PARAPET	06110
6.9	WOOD SHEATHING	06113
6.10	SHEAR WALL BELOW	06114
6.11	WOOD HEADER	06124
6.12	WOOD DECKING	06125
6.13	DUCKBOARD	06125
6.14	GLU-LAM GIRDER	06181
6.15	GLU-LAM BEAM	06181
6.16	WOOD TRUSS	06190

SAMPLE STANDARD KEYNOTES (RESIDENTIAL) continued

ROOF KEYNOTES continued

FINISH CARPENTRY

6.20	WOOD TRIM	06220
6.21	WOOD SOFFIT	06420
6.22	WOOD FASCIA	06420
6.23	WOOD STAIRWORK	06431
6.24	WOOD RAILING	06440
6.25	WOOD GUARDRAIL	06440
6.26	WOOD TRELLIS	06450

DIVISION 7--THERMAL & MOISTURE PROTECTION

WATERPROOFING & DAMPPROOFING

7.1	MEMBRANE WTERPROOFING	07110
7.2	DAMPPROOFING	07150
7.3	VAPOR BARRIER	07190

INSULATION & FIREPROOFING

7.7	RIGID INSULATION	07212
7.8	BATT THERMAL INSULATION	07213
7.9	FOAMED-IN-PLACE INSLATN	07214
7.10	SPRAYED INSULATION	07216
7.11	GRANULAR INSULATION	07216
7.12	FIREPROOFING	07250
7.13	FIRESTOPPING	07270

MEMBRANE ROOFING

7.17	BUILT-UP ROOFING	07510
7.18	PROTECTED MEMBRANE	07550
7.19	REFLECTIVE MEMBRANE	07560

FINISH ROOFING

7.23	ASPHALT SHINGLES	07311
7.24	WD SHINGLES (FIRE HZARD)	07313
7.25	WOOD SHAKES (FIRE HZARD)	07313
7.26	SLATE SHINGLES	07314
7.27	METAL SHINGLES	07316
7.28	CLAY TILES	07321
7.29	CEMENT TILES	07322
7.30	PREFORMED METAL ROOF	07412

SAMPLE STANDARD KEYNOTES (RESIDENTIAL) continued

ROOF KEYNOTES continued

FLASHING & SHEET METAL

7.34	PARAPET CAP FLASHING	07603
7.35	PARAPET FLASHING	07603
7.36	CHIMNEY FLASHING	07608
7.37	SADDLE	07608
7.38	CRICKET	07608
7.39	VALLEY	07609
7.40	HIP	07609
7.41	RIDGE	07609
7.42	SHEET METAL ROOF	07610
7.43	METAL FLASHING	07610
7.44	ROOF GUTTER	07631
7.45	GUTTER SCREEN	07631
7.46	DOWNSPOUT	07632
7.47	RAIN WATER LEADER	07632
7.48	RAIN WATER LEADR STRAINR	07632
7.49	OVERFLOW	07632
7.50	SCUPPER	07633
7.51	RAIN WATER DIVERTER	07634
7.52	EAVE SNOW GUARD	07635
7.53	FLEXIBLE FLASHING	07650
7.54	GRAVEL STOP	07660

ROOF ACCESSORIES

7.58	SKYLIGHT	07810
7.59	SKYLIGHT CURB	07810
7.60	ROOF HATCH	07830
7.61	CABLE SUPPORT & GUY	07871
7.62	JOINT SEALANT	07910
7.63	JOINT FILLER/GASKET	07910

DIVISION 8--DOORS & WINDOWS

8.1	MONITOR SKYLIGHT	08655
10.1	CRAWL SPACE VENT	10200
10.2	LOUVER	10210
10.3	VENT	10210
10.4	EAVE SOFFIT VENT	10210
10.5	SECURITY WINDOW GUARD	10240
10.6	GRILLE	10240
10.7	PREFAB CHIMNEY	10300
10.9	AWNING	10535
10.11	SUN SCREEN	10750

SAMPLE STANDARD KEYNOTES (RESIDENTIAL) continued

ROOF KEYNOTES continued

DIVISION 10--SPECIALTIES

10.1	LOUVER	10210
10.2	ROOF VENT	10230
10.3	RIDGE VENT	10230
10.4	PREFAB CHIMNEY	10300
10.5	CANOPY	10530

DIVISION 13--SPECIAL CONSTRUCTION

13.1	SOLAR COLLECTOR	13610
13.2	PHOTOVLTAIC COLLECTORS	13650

DIVISION 15--MECHANICAL

15.1	EXTERIOR FIRE SPRINKLER	15330
15.2	SOIL VENT STACK	15405
15.3	ROOF DRAIN	15422
15.4	SOLAR WATER HEATER	15431
15.5	AIR HANDLING VENT	15800
15.6	ROOF EXHAUST FAN	15829
15.7	AIR INTAKE	15878

DIVISION 16--ELECTRICAL

16.1	OVERHEAD CABLE ENTRY	16401
16.2	ELECTRICAL SERVICE ENTRY	16420
16.3	LIGHTNING ROD	16601
16.4	TV CABLE ENTRY	16780
16.5	DISH ANTENNA	16781
16.6	SNOW MELTING CABLES	16858

SAMPLE STANDARD KEYNOTES (RESIDENTIAL) continued

FOUNDATION REFERENCE NOTES

1	PROPERTY LINE
2	BENCH MARK
3	SETBACK LINE
4	EASEMENT
5	LINE OF OVERHANG
6	BUILDING LINE
21	EXISTING TO BE REMOVED
22	EXISTING TO REMAIN
23	EXISTING TO BE REPAIRED
31	SLOPE TO DRAIN
32	CONSTRUCTION JOINT

FOUNDATION KEYNOTES

REF #	KEYNOTE	CSI #
	DIVISION 2--SITEWORK	
2.1	GRADING AREA	02210
2.2	EXISTING GRADE	02210
2.3	NEW FINISH GRADE	02210
2.4	LINE OF EXCAVATION	02220
2.5	COMPACTED SUB-GRADE	02220
2.6	COMPACTED FILL	02220
2.7	ROCK SUB-BASE	02230
2.8	EROSION CONTROL	02270
2.11	RODENT CONTROL	02280
2.12	TERMITE CONTROL	02280
2.16	PILES	02360
2.17	CONCRETE RETAINING WALL	02478
2.18	MASONRY RETAINING WALL	02478
2.22	CONCRETE WALK/PAVING	02529
2.23	CONCRETE STEPS	02529
2.27	HOSE BIBB	02640
2.28	WATER SUPPLY MAIN	02665
2.29	GAS MAIN	02685
2.30	GAS METER	02685
2.34	DRAIN	02700
2.35	PERIMETER DRAIN TILE	02710
2.36	CRUSHED-ROCK DRAIN BED	02710
2.37	SUBDRAIN	02710
2.38	SEWER LINE FROM HOUSE	02730

SAMPLE STANDARD KEYNOTES (RESIDENTIAL) continued

FOUNDATION PLAN KEYNOTES continued

DIVISION 3--CONCRETE

3.1	REINFORCING BARS	03210
3.2	WIRE MESH REINFORCING	03220
3.6	CONC. SLAB CONTROL JOINT	03251
3.7	CONSTRUCTION JOINT	03252
3.8	DOWEL CONNECTION BETWEEN SLABS	03253
3.9	CONSTRUCTION JOINT AT GARAGE SLAB AND APRON	03254
3.10	CONCRETE PIER	03303
3.11	CONCRETE GRADE BEAM	03304
3.12	CONCRETE FOOTING	03305
3.13	CRAWL SPACE ACCESS	03305
3.14	CONCRETE FNDATION WALL	03306
3.15	CONCRETE BASEMENT WALL	03307
3.16	CONCRETE SLAB	03308
3.20	FIREPLACE FOUNDATION	03308
3.21	FURNACE SLAB	03308
3.22	SLAB FOOTING	03308
3.23	CONCRETE CURB AT SLAB	03308
3.24	GARAGE PERIMETER SLAB CURB	03308
3.25	GARAGE SLAB SLOPED TO APRON	03308
3.26	GIRDER POCKET	03310
3.27	FOUNDATION WALL LIP FOR MASONRY VENEER	03310
3.28	CONCRETE COLUMN	03317
3.20	PRECAST CONCRETE BEAM	03410

DIVISION 4--MASONRY

4.1	BRICK VENEER	04215
4.2	CONCRETE BLOCK	04220
4.3	CONC. BLOCK FOUNDATION WALL	04229
4.4	STONE VENEER	04450

SAMPLE STANDARD KEYNOTES (RESIDENTIAL) continued

FOUNDATION PLAN KEYNOTES continued

DIVISION 5--METALS

5.1	ANCHOR BOLT	05050
5.2	METAL POST	05120
5.3	STRUCTURAL STEEL	05120
5.4	STEEL JOISTS	05210
5.5	COMPOSITE JOIST SYSTEM	05260
5.6	METAL DECKING	05300
5.7	METAL STUDS	05410
5.8	METAL GRATING	05530

DIVISION 6--WOOD

ROUGH CARPENTRY

6.1	WOOD COLUMN	06101
6.2	WOOD POST	06101
6.3	WOOD GIRDER	06103
6.4	WOOD BEAM	06104
6.5	MUD SILL	06105
6.6	WOOD CAP/POST AT CONC. PIER	06105
6.7	WOOD JOIST	06110
6.8	JOIST BLOCKING	06110
6.9	DOUBLE JOISTS UNDER PARALLEL PARTITIONS	06110
6.10	DOUBLE HEADER JOISTS AT THRU-FLOOR OPENINGS	06110
6.11	WOOD STUDS	06110
6.15	TIMBER CONNECTOR	06111
6.16	FRAMING CLIP	06112
6.17	BEAM HANGER	06113
6.18	JOIST HANGER	06113
6.21	SHEAR WALL	06114
6.22	SHEATHING	06115
6.23	HEADER	06124
6.24	LEDGER	06124
6.25	WOOD DECKING	06125
6.26	WOOD POLE	06133
6.27	WOOD-METAL JOISTS	06150
6.28	GLU-LAM GIRDER	06181
6.29	GLU-LAM BEAM	06181
6.30	WOOD TRUSS	06190

SAMPLE STANDARD KEYNOTES (RESIDENTIAL) continued

FOUNDATION PLAN KEYNOTES continued

DIVISION 7--THERMAL & MOISTURE PROTECTION

7.1	MEMBRANE WATERPRFING	07110
7.2	DAMPPROOFING	07150
7.3	VAPOR BARRIER	07190
7.4	VAPOR BARRIER WITH SAND COVER OVER CRAWL SPACE	07190
7.5	RIGID INSULATION	07212
7.6	PERIMETER THERMAL INSULATION	07212
7.7	BATT THERMAL INSULATION	07213
7.8	FOAMED-IN-PLACE INSULATION	07214
7.9	SPRAYED INSULATION	07216
7.10	GRANULAR INSULATION	07216
7.11	FIREPROOFING	07250
7.12	FIRESTOPPING	07270
7.13	JOINT SEALANT	07910

DIVISION 10--SPECIALTIES

10.1	CRAWL SPACE VENT	10200
10.2	LOUVER	10210
10.3	VENT	10210
10.4	THRU-WALL ACCESS PANEL	10250
10.5	SCUTTLE FROM FLR ABOVE	10250
10.6	RODENTPROOFING	10292
10.7	TERMITE SHIELD	10294
10.8	FIREPLACE ASH PIT AND CLEANOUT	10300

DIVISION 15--MECHANICAL

15.1	HOSE BIBB	15109
15.2	WALL HYDRANT	15109
15.3	GAS METER	15182
15.4	PLUMBING CHASE	15400
15.5	SOIL & WASTE PIPE CHASE	15405
15.6	THRU-WALL SLEEVE FOR SUPPLY PIPE	15410
15.7	THRU-WALL SLEEVE FOR WASTE DRAIN PIPE	15410
15.8	FLOOR DRAIN	15421
15.9	PLUMBING ACCESS PANEL	15423
15.13	WATER HEATER	15450
15.14	FURNACE	15610
15.15	HEAT PUMP	15770
15.16	DUCTWORK	15890
15.17	UNDER-FLOOR PLENUM	15890

SAMPLE STANDARD KEYNOTES (RESIDENTIAL) continued

FOUNDATION PLAN KEYNOTES continued

DIVISION 16--ELECTRICAL

16.1	ELECTRICAL CHASE	16050
16.2	RACEWAYS	16110
16.3	SLEEVE FOR CONDUIT	16110
16.4	ELECTRICAL BURIED CABLE	16400
16.5	ELECTRIC SERVICE ENTRY	16420
16.6	TELEPHONE BURIED CABLE	16740
16.7	TV BURIED CABLE	16780

# # #